Historical Geography and Geosciences

This book series serves as a broad platform for contributions in the field of Historical Geography and related Geoscience areas. The series welcomes proposals on the history and dynamics of place and space and their influence on past, present and future geographies including historical GIS, cartography and mapping, climatology, climate history, meteorology and atmospheric sciences, environmental geography, hydrology, geology, oceanography, water management, instrumentation, geographical traditions, historical geography of urban areas, settlements and landscapes, historical regional studies, history of geography and historic geographers and geoscientists among other topically related areas and other interdisciplinary approaches. Contributions on past (extreme) weather events or natural disasters including regional and global reanalysis studies also fit into the series.

Publishing a broad portfolio of peer-reviewed scientific books Historical Geography and Geosciences contains research monographs, edited volumes, advanced and undergraduate level textbooks, as well as conference proceedings. This series appeals to scientists, practitioners and students in the fields of geography and history as well as related disciplines, with exceptional titles that are attractive to a popular science audience.

If you are interested in contributing to this book series, please contact the Publisher.

Jacobo García-Álvarez ·
André Reyes Novaes
Editors

Mapping the Contested Line

Cartography and Border Disputes in the Iberian Peninsula, North Africa and Latin America (18–20th Centuries)

Springer

Editors
Jacobo García-Álvarez
Department of Humanities: History, Geography and Art
Universidad Carlos III de Madrid
Getafe, Spain

André Reyes Novaes
Department of Human Geography
Universidade do Estado do Rio de Janeiro
Maracanã, Brazil

ISSN 2520-1379 ISSN 2520-1387 (electronic)
Historical Geography and Geosciences
ISBN 978-3-032-29888-1 ISBN 978-3-032-29889-8 (eBook)
https://doi.org/10.1007/978-3-032-29889-8

This work was supported by Spanish State Research Agency.

This Springer imprint is published by the registered company Springer Nature Switzerland AG
The registered company address is: Gewerbestrasse 11, 6330 Cham, Switzerland

Preface

Mapping the Contested Line brings together a number of case studies on the relationships between cartography and border disputes, using map history to frame the construction of nation-states and colonial territories in the Iberian Peninsula, North Africa and Latin America from the eighteenth to the twentieth century. The book includes eight contributions by scholars from Spain, France, Portugal, Argentina, Uruguay and Brazil, and has two main parts. Part I (Chaps. 2–5) focuses on the use and contribution of maps to the processes of negotiation and boundary fixation on disputed borders, both in diplomatic forums and in the materialisation of the border on the ground, whereas Part II (Chaps. 6–8) delves into the circulation of maps used by boundary commissions, as well as into how the political purposes of cartography and the audiences to whom maps were addressed and communicated conditioned the very representation of borders. In addition to these parts, the introductory chapter provides an overview of recent scholarship about the links between maps and border issues, both at a general level and in the countries and regions specifically considered in this volume. The case studies included in this volume cover a broad and varied spectrum, whether from a chronological, geopolitical or landscape perspective, ranging from the mid-eighteenth century to the mid-twentieth century, from colonial to post-colonial spaces, and from international rivers and border wetlands of the Iberian Peninsula and South America to the rocky and sandy deserts of Western Sahara.

Far from just a collection of case studies with specific and empirical interests, this book aims to contribute to the whole field of map history. The challenging use of maps to establish static and permanent lines in fluid physical and human environments, such as rivers, mountains and deserts, and the use of maps as discourses to convince and negotiate borders in diplomatic, military and popular realms could showcase how maps are used in practices, overcoming idealised and decontextualised approaches in map history. Likewise, examining the circulation and communication of cartographic images involved in border disputes can contribute to developing a critical and transnational approach to border mapping, as well as to putting into question certain nationalist or colonialist narratives surrounding the production of geographical knowledge. By addressing the relationships between mapping and contested borders in different historical and geographical contexts, this collective contribution, based primarily on archival sources, seeks to approach a series of methodological challenges and contemporary controversies that encompass the making of the history of cartography.

The edition of this book and the writing of all chapters (except Chap. 6) have been supported by Grant Number PID2020-114088GB-I00/AEI/10.13039/501100011033, funded by the Spanish State Research Agency.

Getafe, Spain — Jacobo García-Álvarez
Rio de Janeiro, Brazil — André Reyes Novares

Contents

Editors and Contributors

About the Editors

Jacobo García-Álvarez is Professor of Human Geography at the Universidad Carlos III de Madrid (UC3M), which he joined in 2003. His research work has been focused on the fields of history of geography, regional geography, historical geography and political geography and geopolitics, with special attention to politico-administrative divisions, boundary-making processes, border conflicts, and territorial identities and ideologies in Spain and Latin America. He has lectured in several universities and research centres in Europe, Latin America, and the USA. He chaired the Commission on the History of Geography at the International Geographical Union and the International Union of History and Philosophy of Science and Technology from 2008 to 2016. He has coordinated the History of Geography, Historical Geography and Political Geography Research Group in the Julio Caro-Baroja Institute of Historiography at the UC3M since 2012.

André Reyes Novaes is Associate Professor at the Universidade do Estado do Rio de Janeiro, Brazil. He is an Honorary Research Associate at Royal Holloway, University of London. He was a visiting Scholar at the University of Nottingham and a Fellow at the John Carter Brown Library Fellow. He is a member of the commission on the history of geography at the International Geographical Union (IGU), and he is one of the editors of the book series Geographers: Biobibliographical Studies. He edits journals on historical and cultural geography, such as the *Journal of Historical Geography* and the *Espaço e Cultura* Journal, and contribute in the editorial board of journals on the history of cartography and the history of geography, such as the *Imago Mundi* journal and the Brazilian journal *Terra Brasilis.* His research interests include visual methods, history of cartography, popular geopolitics and history of South American borders.

Contributors

Jacobo García-Álvarez Department of Humanities: History, Geography and Art, Universidad Carlos III de Madrid, Getafe, Madrid, Spain

Luís Miguel Moreira Geography Department, Institute of Social Sciences, University of Minho, Guimarães, Portugal;

Centre of Geographical Studies, Institute of Geography and Spatial Planning, University of Lisbon, Lisbon, Portugal

André Reyes Novaes Department of Human Geography, Universidade do Estado do Rio de Janeiro, Maracanã, Rio de Janeiro, Brazil

Sergio Nunes-Pereira Fluminense Federal University, Niterói, Brazil

Jean-Yves Puyo Laboratoire TREE (UMR 6031 CNRS), University of Pau and Pays de L'Adour, Pau, France

Ana Raya-Collado Department of Humanities: History, Geography and Art, Universidad Carlos III de Madrid, Getafe, Madrid, Spain

Lucía Rodríguez Arrillaga Department of American History, Universidad de la República, Montevideo, Uruguay

Perla Zusman CONICET and University of Buenos Aires, Buenos Aires, Argentina

1 Contested Lines and Mapping Practices: An Introduction to the Relationships Between Cartography and Border Disputes in the Iberian Peninsula, North Africa and Latin America

Jacobo García-Álvarez and André Reyes Novaes

Abstract

This chapter examines the relationships between cartography and border disputes across the Iberian Peninsula, North Africa and Latin America. It discusses the complex role of maps in defining fixed political boundaries within fluid physical and human environments and their use as persuasive tools in diplomatic, military and popular realms. By emphasizing mapping practices, the chapter aligns with methodological trends that challenge idealized and decontextualized histories of cartography. The persistent notion of cartography as a unified, progressive and neutral enterprise often parallels the ideal of the nation-state as a clearly bounded and sovereign entity. In contrast, this study approaches maps as geopolitical instruments that sustain and legitimize nationalist projects. The chapter first outlines historical connections between maps and border disputes, before exploring how these dynamics played out in the Iberian, North African and Latin American contexts. It concludes by introducing the volume's case studies, which collectively illustrate diverse temporal, political and geographical perspectives ranging from the mid-eighteenth to the mid-twentieth century. By approaching a series of methodological challenges that encompass the study of mapmaking and boundary-making, the volume seeks to contribute to ongoing interdisciplinary debates in fields such as map history, historical geography and history of science.

Keywords

Cartography · Geopolitics · Diplomacy · Border maps · Border disputes · Boundary-making · Nation-state · Iberian Peninsula · North Africa · Latin America

1.1 Maps and Boundaries

During the last thirty years, an increasing number of studies have examined the relation-

This work has been supported by the Spanish State Research Agency under Grant Number PID2020-114088GB-I00/AEI/10.13039/501100011033, as well as by the Universidad Carlos III de Madrid under Grant Number 2012/00017/003.

J. García-Álvarez (✉)
Department of Humanities: History, Geography and Art, Universidad Carlos III de Madrid, Getafe, Madrid, Spain
e-mail: jacobo.garcia@uc3m.es

A. R. Novaes
Department of Human Geography, Universidade do Estado do Rio de Janeiro, Maracanã, Rio de Janeiro, Brazil
e-mail: andre.reyes.novaes@uerj.br

J. García-Álvarez and A. Reyes Novaes (eds.), *Mapping the Contested Line*, Historical Geography and Geosciences, https://doi.org/10.1007/978-3-032-29889-8_1

ships between maps and conflicts developed in the borders or because of borders, considering cartography as an important and constitutive tool in nation-states' building processes (Anderson 1983; Newman 2006; Edney 1997). Nation-states, like empires, have been considered the most relevant political contexts in the history of cartography, because of their ability to depict territorial contours and popularize national borders (Harley 1988). To approach the relationship between maps and borders, we would like to begin by addressing an act that seems simple yet powerful: drawing a line.

In order to open his book on the history of space and the emergence of a 'cartographic reason', Pickles (2004) revisits a text by Gunnar Olsson in which the Swedish geographer famously asked: 'What is geography if it is not the drawing and interpreting of a line?' (Olsson 1992). Drawing a line creates a new object and produces social effects. It is a fundamentally geographical act through which territories are fixed and identities inscribed (Pickles 2004, p. 3; Puente-Lozano and García-Álvarez 2022). Influenced by Olsson's notion of 'lines of power', Pickles challenged approaches that define maps as merely bidimensional representations of pre-existing territorial features. Border maps are perhaps the most evident example supporting the claim that mappings precede the territories they purport to represent. Maps do not simply depict territory; rather, they provide the 'very condition of possibility of the worlds we inhabit and the subjects we become' (Pickles 2004, p. 5).

The drawing of lines and the bounding of objects have been key practices in producing distinctions, taxonomies and hierarchies. This is a central theme of the present book, which is concerned with the entanglement of mapmaking and boundary-making. Inscribing lines on a surface is not a purely technical act; it requires the layering of different forms of inscription in order to negotiate and demarcate boundaries. Drawing lines on paper is not merely the outcome of scientific expeditions or field practices. It is shaped by embodied experiences, environmental challenges and power relations. By treating lines traced on the ground and lines drawn on paper as interlocking practices, this volume demonstrates how cartographic reason has coded subjects and produced territorial identities. Based on original and detailed archival research, the chapters assembled here engage in different ways with the processes that Olsson evocatively associated with the 'finger' and the 'eye'. These metaphors informed Pickles's (2004) reflections on the emergence of a cartographic reason and have also inspired the conceptual organization of this book.

The first part of the book engages with Olsson's metaphor of the 'finger', approaching pointing and naming as categorizing practices that 'stabilize a particular meaning within a world of possible meanings' (Pickles 2004). To point to a place and to name it has been a fundamental strategy in the extension of imperial and national control over both physical and human environments (Williamson 2023). The use of cartography in the diplomatic negotiation of border disputes and in the delimitation of contested frontiers exemplifies this indexical role of maps: they fix lines that are mobile and unstable on the ground. In such cases, cartographic reason operates by translating fluid and dynamic features (rivers, deserts, settlements) into stable, fixed and ideally immutable lines. The arguments mobilized to justify a territorial boundary (natural frontiers, *uti possidetis*, or prior treaties) may differ substantially, yet they share a common aim: to inscribe on the map what is fluid, dynamic, or even unknown in the terrain.

The second part of the book turns to Olsson's metaphor of the 'eye', considering vision as a set of technologies and practices crucial to the circulation and dissemination of worldviews. The trajectories of border maps demonstrate how their meanings shift according to different 'eyes', foregrounding the mobility of maps and the variability of their reception. Depending on their contexts of circulation, maps may be disputed, shared, archived, stolen or ignored, rendering their symbolic value deeply contextual and often controversial. As instruments for translating the world into exhibitionary form, maps have been central to colonial and national economies of display and demarcation (Gregory 1994). When the 'eye' apprehends a drawn line, the world appears domesticated and ordered, producing dualistic distinctions—native and foreigner, Westerner and Easterner, us and them (Pickles 2004, p. 5).

These dualisms have been increasingly challenged over the past decades, as scholars have shown growing interest in studying asymmetrical processes of encounter, entanglement and connection in mapmaking practices (Subrahmanyam 1997; Raj 2007; Craib 2009; Novaes 2025). The emphasis on the circulation of knowledge advanced by historians of science has helped to relocate so-called modern science on a global scale (Jöns et al. 2017), encouraging map historians to understand "modern" cartography as the product of the combination, translation and appropriation of multiple knowledge traditions (Raj, 2007). The histories of the lines drawn to define and legitimize political, administrative, identity-based and territorial divisions are clearly not exclusively European. Spatial and representational practices aimed at delimiting otherness have emerged in distinct ways across diverse regions and historical contexts. In China, for example, historical atlases produced between 900 and 1005 depicted the empire's borders, emphasizing territorial gains and losses negotiated or fought over with groups regarded as 'non-Chinese' (Standen 1994; Yee 1994; Black 2005). In India, rulers such as the Mughals and Marathas employed maps during the eighteenth century to delineate boundaries, administer cultivated lands and organize revenue collection (Gogate and Arunachalam 1998).

The European endeavour intensified between the fifteenth and sixteenth centuries, when maps were increasingly used to record and negotiate borders, particularly in the drafting of treaties concerning both internal and external frontiers. Notable examples include the Treaty of Ardres between France and England (1546) and the Treaty of Tordesillas between Spain and Portugal (1494) (Black 2005, 40). The printing of maps began to spread across Europe from the 1470s onward (Woodward 1975), but it was especially during the sixteenth and seventeenth centuries that print culture fostered what has been described as a 'public world of maps' (Black 2005, p. 24). This development inaugurated a more frequent use of maps as symbolic and communicative tools for identity formation. From this period forward, an expanding audience of 'eyes' could engage with cartographic representations.

Despite earlier examples of maps depicting borders in different regions and periods, the eighteenth century marked a crucial milestone in the use of cartography for the diplomatic resolution of conflicts and territorial claims, both in Europe and in its colonial domains, and later in the states that emerged through decolonization (Black 1994; Prescott and Triggs 2008; Donaldson 2008; Seegel 2012; Warmoes et al. 2016; García-Álvarez and Puente-Lozano 2022; Górny 2022). It was also during this period that maps enhanced their capacity to disseminate the territorial contours of emerging nation-states within the imagination of their citizens, transforming cartography, in Benedict Anderson's well-known formulation, into a powerful icon or logo of the national idea (Anderson 1983).

The growing use of maps in the negotiation and communication of territorial claims eventually prompted institutional initiatives to collect, reproduce and generate cartographic materials. For example, the French Ministry of Foreign Affairs established a geographic section in 1772 and, in 1780, acquired the geographer Jean-Baptiste d'Anville's collection of approximately 10,000 maps. This acquisition illustrates how 'diplomats and politicians increasingly turned to maps in crises and times of war' (Black 2005, p. 41). Many of the archives consulted in the preparation of the chapters in this volume were likewise created in the eighteenth century amid efforts to gather and systematize cartographic materials. Institutions in both Latin America and Europe began assembling map collections to support the negotiation and demarcation of national borders. These repositories constitute fundamental sources for studying the history of border making and its visual manifestations. As Black (2005) aptly observed, border making is a complex and multilayered process in which maps have played diverse and shifting roles, each of which required maps to mobilize different forms, scales, languages and rhetorical strategies. Such diversity challenges attempts to define border maps through a narrow set of practices or fixed patterns of production, circulation and consumption. In seeking an open criterion to group border maps, we aim to avoid defining them as a specific 'genre' or 'mode'. A map genre presupposes that maps can be classified into relatively stable types according to thematic content, linguistic style and compositional structure (Lois 2015), while a cartographic mode identifies recurring patterns of production, circulation and use.

In his effort to provide conceptual tools for understanding the various ways people have mapped their worlds, Matthew Edney proposed fourteen mapping modes. Border maps appear as the first mode within the group he terms 'Modern, state-driven mappings'. In this framework, boundary maps are necessarily state practices, typically covering 'relatively narrow areas over geographical distances along a border or frontier between states' (Edney 2019, p. 33). Although many of the maps examined in this volume fit this criterion, a closer engagement with practices of border negotiation, delimitation and circulation suggests that it is ultimately too restrictive. Archival research reveals that border making is a multifaceted process involving maps produced at different scales, in diverse languages, and according to varied visual conventions.

When the lines dividing sovereign territories are controversial and contested, cartography has been mobilized in multiple forms to render them visible, thereby generating, sustaining and disseminating boundaries in both national and colonial contexts. As Gerald Blake has well summarized,

> Maps are important in international boundary disputes in four major ways: First, they may be contributory factors, or even causes of the dispute; second, they may be used by the parties to promote their own position in a dispute; third, they are frequently among the most useful tools in dispute resolution; and, fourth, they may be used to illustrate the judgement of mediation or arbitration (Blake 1995, p. 48).

In colonial settings in particular, the persistence of *terrae incognitae* and the scientific uncertainties associated with certain frontier zones gave rise to specific kinds of maps, often filled with conjecture and guesswork. Such materials resist classification within Edney's relatively narrow definition of 'boundary maps'. It is important to note, however, that Edney himself emphasizes the fluidity of his categories, repeatedly stressing that cartographic modes are open and constantly in flux. The ongoing exchange and interaction among mapping modes over time, he argues, constitute a central object of study for historians of cartography (Edney 2019).

In this volume, we define border maps broadly as any cartographic image involved in the depiction, negotiation, demarcation or circulation of boundaries, regardless of scale, language, representational practice or function. If an image has been used to draw material or symbolic lines between individuals and territories, we consider it a border map. When examined in light of their production and consumption practices, such images reveal a wide range of techniques and visual languages, underscoring how the presumed intrinsic relationship between the state and cartography is, in fact, an idealized narrative constructed over the course of the eighteenth century.

Maps have functioned as geopolitical tools and, as such, have contributed to the dissemination of nationalist projects intertwined with expansionist, irredentist or separatist agendas. These persuasive or propagandistic uses of cartography are well documented. Yet maps produced in the field, through the application of instruments and measurement techniques to survey terrain and resolve boundary disputes, could likewise become 'weapons', both in diplomatic arenas and in the broader public sphere (Herb 1997; Leuenbeger and Schnell 2020; Kaya 2020).

The ideal of cartography, understood as a belief system as well as a discursive and historiographical construct, has consistently developed alongside the ideal of the nation-state, conceived as an exclusive, linear and sovereign territorial formation. Recent scholarship has brought the histories of cartography and political thought into dialogue, demonstrating how maps have been central to the construction of national territorialities, the consolidation of modern sovereignty, and the generalization of linear borders (Branch 2013; Biggs 1999; Strandsberg 2010; Altíc 2022; Debarbieux 2022; Goettlich 2025).

According to the authors who have explored this research agenda, the cartographic endeavours initiated in Europe since the fifteenth century not only occurred in parallel with the formation of modern states, providing a key technological instrument to represent, claim or negotiate territories. More than that, those endeavours also changed the ways of understanding political authority, enabling the change from a jurisdictional and patrimonial conception of sovereignty, typical of feudal society, to a territorial one, understood as a crucial characteristic of the nation-state in the current world. As pointed out by Branch (2013), the modern state could be framed as a 'cartographic state' because its territoriality had a 'geometric or cartographic nature', according to which states are delimited by linear boundaries enclosing contiguous and compact territories and clearly dividing areas of exclusive sovereignty, although they may include disputed lands.

The growing entanglement between mapmaking and boundary-making in the eighteenth century was a global phenomenon that must be understood in light of transnational circulations of knowledge. During this period, unprecedented efforts at boundary mapping were undertaken not only in Europe but in many other parts of the world. Although borders had existed for centuries, borderlines, understood as precise cartographic divisions superimposing exclusive imperial sovereignty onto continuous territories, were largely an eighteenth innovation that reshaped mapping practices worldwide (Erbig 2020, p. 3).

In Europe, boundary mapping shifted from fluid and locally negotiated demarcations towards more precise, state-driven cartography, prompted by military, administrative and fiscal demands (Edney and Pedley 2019). In China, the Qing Empire invested heavily in mapping its northwestern frontier in competition with the expanding Russian Empire (Forêt 2000). In Africa, new cartographic projects sought to fill blank spaces and eliminate mythological references, reinforcing colonial ambitions to draw new lines across the continent (Furtado 2017). In Latin America, negotiations between the Spanish and Portuguese empires culminated in the Treaty of Madrid (1750) and the Treaty of San Ildefonso (1777). These agreements generated an unprecedented practice: the dispatch of joint boundary commissions tasked with traversing vast territories and surveying borderlines directly on the ground (Erbig 2020).

All these transformations in mapping practices justify the temporal scope of this book, which explores case studies of border mapping from the eighteenth to the twentieth century. Although the relationship between mapmaking and the emergence of states and empires was a wide-ranging global phenomenon, this volume adopts a more focused geographical lens, concentrating on the Iberian Peninsula, North Africa and Latin America. These regions share deeply intertwined histories, and their long-standing interdependence, marked by asymmetrical power relations, provides a particularly productive framework for examining the circulation, accommodation and appropriation of cartographic knowledge between colonies and metropolises.

1.2 Cartography and Border Disputes in the Iberian Peninsula, North Africa and Latin America

The study of border disputes during the eighteenth and twentieth centuries in the regions covered by this book has given rise to an extensive bibliography (for the Iberian states and empires see, for example, Cordero 1960; Herzog 2015; García-Álvarez and Puente-Lozano 2017;

Melón et al. 2019; García-Hernán et al. 2023; for postcolonial Latin America, Foucher 1991; Parodi 2002; Prescott and Triggs 2008: 233–264, Infante 2013; Cairo 2023: 209–268; and Zusman 2025). During the colonial period, the definition of the borders of the Spanish and Portuguese monarchies in Southern America involved several disputes and boundary treaties both between the two Iberian kingdoms and with other European powers, such as France.

These colonial-type border disputes were also reproduced in the nineteenth and twentieth centuries in other geographic areas with an Iberian presence, such as Asia and Africa. In parallel with these controversies, both the absolute monarchy and, above all, the liberal governments in Spain since the 1830s promoted different Treaties with France (1856–1868) and Portugal (1864, 1893 and 1926) to delimit their land borders on the Iberian Peninsula according to the aforementioned 'modern' criteria. Within this context, cartographic representations were used to assert their exclusive sovereignty over a compact state space defined by linear, clear and continuous border limits. From the nineteenth century onwards, this model of state territoriality was exported from Europe to other continents and regions due to imperialism and the independence processes of the ancient colonies, starting with the Latin American ones.

In these terms, the principle of *uti possidetis iuris*, proclaimed by the Congress of Angostura (1819) as a general criterion with which to guide the formation of new states emancipated from Spanish dominion, did not entail in any way the fossilization of colonial limits. According to Michel Foucher's calculation in his main contribution to the question (Foucher 1991), only approximately 27% of the total length of the current land borders of Latin American countries is of colonial origin (i.e. it comes either from inter-imperial borders or from intra-imperial administrative limits). Another 60% was delimited in the century that followed the processes of independence, so that in 1914, when the Panama Canal was inaugurated, around 90% of the current border lines of Latin American countries had been completed. Regarding the explanatory factors of the current border limits, and apart from that 27% of colonial origin, some 30% of their total length was the result of wars, another 17.5% was imposed by one of the parties (whether due to internal or external interests), 17% come from treaties negotiated between countries that emerged from independence and 6.6% resulted from arbitration.

The discontinuity of colonial and national borders identified by Foucher highlights the importance of studying the constitution of borders in the nineteenth and twentieth centuries, when decolonization processes advanced and negotiations no longer depended on a position centred on Europe. Within the scope of the history of cartography, as well as in other fields concerned with the political dimensions of maps (such as the history of geography and territorial ideas and knowledge, historical geography, political geography and geopolitics), the study of relationships between borders and border disputes in the Iberian states and Latin America has given rise to different lines of research, two of which are worth highlighting here (a recent state of the art on this literature can be seen in García-Álvarez and Puente-Lozano 2023).

A first line would group those works that, through the exhaustive exploitation of unpublished archival sources, have allowed us to better understand the role of certain individuals, as well as certain institutions, in the tasks of delimiting borders associated with the preparation or execution of modern Boundary Treaties. These studies often focus on sentences and arbitration reports produced when disputes are resolved by legal procedures, referring both to colonial territories (in which, in many cases, delimitation and demarcation processes were inextricably mixed with scientific exploration) and to metropolitan and postcolonial areas. In the case of Spain and the Spanish Empire, as well as for the countries that emerged from Latin American independence, this type of work can demonstrate the relevance of some army corps, such as naval officers, military engineers or general staff members, with their respective training academies and documentation centres. Likewise, an important number of studies have investigated the role of joint or bilateral boundary commissions, a type of institution that has given rise to important studies within and outside of Spain, some of them comparative (Lucena 1993; Ferreira 2001; Penhos 2005; Urteaga 2007; Gesteira 2014; Erbig 2015; García-Álvarez and Puente-Lozano 2015, 2022; García-Álvarez and Puyo 2015, 2019; Lois 2019; Manso 2016; Haller and Tapia 2023).

A second line of investigations has focused on the use of geography and geopolitics in the service of nationalist territorial claims, both at the level of academic discourses and at the level of the so-called popular geographies (or 'geopolitics'), which are fundamental for the dissemination and national socialization of these claims. Along this line, for example, the role of persuasive or propagandistic cartography and, above all, the role of school history and geography teaching in the legitimation and social reproduction of certain territorial aspirations and border disputes has been extensively explored (see, for instance, Escudé 1989; Romero 2004; Reguera 2007; Lois 2012; García-Álvarez 2013; Herzog 2017; Del Castillo 2018; Novaes 2019; Zárate 2020; Zusman 2025).

Both lines of inquiry are essential for understanding the circulation of cartographic knowledge across multiple directions and scales from the eighteenth century onward. A close examination of border treaties and boundary commissions reveals how lines were negotiated, contested and materially produced in practice. At the same time, attention to the circulation and persuasive power of maps enables us to assess their broader social and political effects. Experiences of mapping and the linearization of borders were undoubtedly intertwined with European colonial expansion. Yet this does not mean that mapping practices were consistently formulated and tested in the Iberian Peninsula before being implemented in colonial settings. Cartographic knowledge moved along multidirectional routes. A dense flow of spatial information connected Iberian metropolises with their colonies, and later with the independent nations of Latin America and territories in North Africa. In many instances, colonial experiences preceded and informed practices subsequently adopted in metropolitan contexts. Our spatial scope is designed precisely to trace these diverse circulations of knowledge, situating the production of maps and borders within interconnected colonial and colonialist frameworks.

1.3 Negotiating, Fixing, Circulating and Communicating the Contested Line: A Summary of the Chapters

In addition to this introductory chapter, this book brings together seven contributions representative of the research lines outlined above, as well as of other lines of inquiry and themes relevant to the international historiography of cartography. The choice of the topics and case studies analysed in these chapters is based on several reasons. Firstly, six of those case studies are part of a collective research project focused on border conflicts in Spain and Latin America in the nineteenth and twentieth centuries, which includes the analysis of both the Spanish-Portuguese boundaries of the Iberian Peninsula and some African or Latin American border areas that were Spanish possessions in the colonial past. Although the selected cases are diverse in time and space, they allow us to approach and examine, from an Iberian comparative perspective, the theories and practices of border delimitation developed by both the Spanish and Portuguese empires, as well as the role played by maps in the processes of boundary-making and in the conflicts associated with them.

In addition, the essays collected in this volume demonstrate how these theories and practices were inherited, adapted and transformed by the governments of the Latin American states that emerged from the independence processes within the former Spanish and Portuguese empires. As the chapters show, there are significant continuities in the institutions, practices and principles of border diplomacy operating under both colonial and national regimes. Among these continuities are the prominence of Joint Boundary Commissions as key instruments of delimitation and conflict resolution; the decisive role played within them by military corps trained in cartographic techniques; and the invocation of the principle of *uti possidetis de facto* (as opposed to *uti possidetis iuris*). These elements were appropriated and recalibrated by the nineteenth-century Latin American nation-states. The principle of *uti possidetis de facto*, in particular, had earlier been advanced by Portuguese governments and was subsequently mobilized by Brazilian authorities to legitimize sovereignty over extensive areas of the American continent unilaterally occupied beyond the terms of boundary treaties concluded before the second half of the eighteenth century. More broadly, the legal arguments, diplomatic mechanisms and cartographic procedures deployed to define borders and manage disputes reveal a marked continuity between colonial and national contexts.

Finally, the selection of cases illustrates some of the main uses of cartography in border disputes. Covering the different map roles in international boundary disputes identified by Blake (1995), the chapters showcase how maps could be a cause of dispute, a means of promoting the positions and claims of each party, or a tool for collaborating in the resolution of disputes, either by helping to demonstrate the problems at the root of the dispute or by assisting in proposing solutions. In a sense, the chapters assembled here function as a broad inventory of the diverse maps and mapping practices mobilized in the processes through which borders were produced.

As briefly presented in the beginning of this chapter, the seven contributions that follow this introductory essay, have been divided into two parts, each of which addresses some central aspects of the relationship between cartography and border disputes. The first part of the book (Chaps. 2 to 5) focus on the use of cartography in the diplomatic negotiation of border disputes and in the fixing of contested borders, whereas the second part (Chaps. 6 to 8) is dedicated to studying the circulation and communication of border maps, exploring how those maps can change their meanings and roles in relation to their mobilities and receptions.

In Chap. 2, Jean-Yves Puyo addresses the weight of cartographic documents in the arbitration of the Franco-Brazilian dispute over the border between French Guiana and the Brazilian province of Amapá, issued by the Swiss Federal Council in 1900. The controversy actually dated back to the Treaty of Utrecht (1713), which divided French and Portuguese possessions in this region through a series of articles that the Brazilian and French governments interpreted differently. Drawing on French and Swiss diplomatic archival documentation, the author analyses the numerous maps presented by both delegations in defence of their respective positions. Despite the considerable cartographic compilation effort carried out during the process, and the use of geographical arguments supported by that cartography, Puyo concludes that this documentation had a very limited impact on the final outcome of the arbitration, which was favourable to the Brazilian positions, based on the principle of *uti possidetis de facto*.

Chapters 3 and 4 reflect on the problems of delimiting river borders and examine the use of cartography associated with these processes. Each of these chapters addresses a significant case study in this regard: while, in Chap. 4, Jacobo García-Álvarez deals with the Spanish-Portuguese border along the Minho River, which was delimited between 1855 and 1898, in Chap. 5 Perla Zusman and Sergio Nunes focus on the Argentine-Brazilian boundary in the Misiones/Palmas region, along the Uruguay, Pepiri-Guazu, San Antonio and Iguazu rivers, decided in 1895 by an arbitration award issued by the U.S. President Grover Cleveland and demarcated on the ground between 1901 and 1905.

As discussed in both contributions, despite the frequency with which rivers have served as international boundaries, such kind of borders pose significant legal and geopolitical problems due to the fact that water is a fluid and dynamic phenomenon, while the concept of territorial sovereignty entails the ideals of fixity and permanence, as well as because of the difficulties of managing and administering these transboundary areas in a unitary way. The case studies discussed in the chapters illustrate some of the main challenges involved in fixing these 'moving lines', such as determining sovereignty over river islands and waterfalls, defining a dividing boundary in watercourses affected by high variability, or regulating certain human uses and constructions whose lack of control has the potential to cause harmful repercussions. Likewise, both chapters delve into the role played by cartography in illustrating and interpreting the problems associated with river borders; the technical difficulties inherent in the cartographic surveying of watercourses; the strategies used to map the changing landscape of the river; the production and use of maps to support the competing claims of the litigating parties; and, finally, the contribution of cartography to the implementation of certain territorial planning measures.

To conclude Part I, in Chap. 5 Ana Raya-Collado and Jacobo García-Álvarez reflect on the role of cartography in the delimitation and demarcation of colonial spaces associated with desert environments, taking the Spanish Western Sahara as an illustrative case study. The first part of the chapter deals with the particularities of boundary-making processes in the aforementioned environments, with special attention to the frequent use of geometric boundaries and to the role of maps in the appropriation and colonization of these spaces, as well as in the boundary disputes linked to them. The second part of the chapter focuses on the example of Western or Spanish Sahara, whose borders were defined in the Spanish-French conventions of 1900, 1904 and 1912, but were neither delimited nor demarcated until the 1950s. As the authors argue, the cartography produced during this period was key in the process of geometrization and administrative purification of the borders of this mostly desert region, facilitating the transition from a zonal frontier, such as the one that separated the Spanish and French colonial administrations for several decades, to a linear boundary, linked to the European modern ideas of sovereignty and territoriality and superimposed over the spatial conceptions and patterns of the native nomadic tribes.

The second part of the book (Chaps. 6 to 8), dedicated to studying the circulation and communication of border maps, opens with the Chap. 6, in which Lucía Rodríguez-Arrillaga explores the circulation and validation of maps in the delimitation of the Río de la Plata in the mid-eighteenth century. The exchange and validation of maps in a trans-imperial context is an interesting topic that could challenge static and internalist approaches in map history. Tracing the circulation and mobilities of printed and manuscript maps, exchanged by officials of the Spanish and Portuguese crowns, the study reveals how maps functioned as tools of political validation and epistemic authority. Maps printed in Paris (1748), for instance, made by mapmakers who never travelled to South America, such as D'Anville, were often taken more seriously than locally produced maps, made by native people or local agents.

A similar difference of cartographic authority is also raised in Chap. 7 by André Reyes Novaes. The chapter combines archives in collections at the Royal Geographical Society, in the UK, and the Museu Historico e Diplomático do Itamaraty, in Brazil, to approach how maps acquire distinct meanings and functions depending on their archival contexts and spaces of circulation. By approaching the practices of mapping and demarcation along the Verde River—on the Brazilian-Bolivian frontier between 1908 and 1909—the chapter moves beyond national frameworks to uncover multiple narratives and standpoints in the production of geographical knowledge. The archives of two military engineers who served respectively as subchief and chief of the Bolivian and Brazilian boundary commissions, Percy Harrison Fawcett and José Cândido Guillobel, allowed the author to situate cartographic credibility in specific communities. While Fawcett's maps were received as a novelty in the mapping of *terrae incognitae* in his presentations at the Royal Geographical Society and his publications at *The Geographical Journal,* they had little impact among diplomats who were negotiating new boundaries in

South America. The chapter calls for an archival combining across languages, institutions and locations as a tool to construct a transnational perspective on border mapping practices.

Finally, in Chap. 8, Luís Miguel Moreira explores the production of topographical maps on the border between Spain and Portugal in the Iberian Peninsula during the nineteenth century. In order to remove legal ambiguities inherited from the Middle Ages, the international boundary commission responsible for implementing the 1864 Boundary Treaty produced maps in different scales, and the chapter compares these cartographies, which were produced for political and diplomatic negotiations, with topographic maps made to circulate in military and civilian circles. The comparison of these maps showcases the role of cartography in the state practices aimed at ruling border areas, as well as the involvement of some of those border representations in nationalist narratives, marking the distinction between the material, cartographic evidence for the Portuguese "us" and the vast cartographic silence regarding the Spanish 'other'.

The case studies included in this volume cover a broad and varied spectrum, whether from a chronological, geopolitical or landscape perspective, ranging from the mid-eighteenth century to the mid-twentieth century, from colonial to postcolonial spaces, and from international rivers and border wetlands of the Iberian Peninsula and South America to the rocky and sandy deserts of Western Sahara. But, far from just a collection of case studies with specific and empirical interests, this book aims to contribute to the whole field of map history. The focus on maps, actors and institutions is quite explicit in all chapters, and it is clear that the investigations conducted could raise different methodological and epistemological debates in fields such as map history, historical geography and the history of science.

The challenging use of maps to establish static and permanent lines in fluid physical and human environments, such as rivers, mountains and deserts, and the use of maps as discourses to convince and negotiate borders in diplomatic, military and popular realms could showcase how maps are used in practices, overcoming idealized and decontextualized approaches in map history. Likewise, examining the circulation of cartographic images involved in border disputes can contribute to developing a critical and transnational approach to border mapping, as well as to putting into question certain nationalist or colonialist narratives surrounding the production of geographical knowledge. By addressing the relationships between mapping and contested borders in different historical and geographical contexts, this collective contribution, based primarily on archival sources, seeks to approach a series of methodological challenges and contemporary controversies that encompass mapmaking and boundary-making.

References

Anderson B (1983) Imagined communities. Reflections on the origin and spread of nationalism. Verso, London

Altíc M (2022) From borderlands to boundary lines: mapping on the edges of the Ottoman Empire. Imago Mundi 74(2):189–216. https://doi.org/10.1080/03085694.2022.213052

Biggs M (1999) Putting the state on the map: cartography, territory, and European state formation. Comp Stud Soc Hist 4(2):374–411

Black J (1994) Boundaries and conflict. International relations in ancient régime Europe. In: Grundy-Warr C (ed) World Boundaries. Volume 3: Eurasia. Routledge, London and New York, pp 19–54

Black J (2005) Mapas e História. Construindo Imagens do Passado. Edusc, Baurú

Blake G (1995) The depiction of international boundaries in topographic maps. IBRU Bound Secur Bull 3(1):44–50

Branch J (2013) The cartographic state: maps, territory, and the origins of sovereignty. Cambridge University Press, Cambridge

Cairo H (2023) Geopolítica de América Latina y el Caribe. Síntesis, Madrid

Cordero JM (1960) Fronteras hispánicas. Geografía e historia. Diplomacia y administración. Instituto de Estudios Políticos, Madrid

Debarbieux B (2022) The cartographic factory of modern states. In: Debabieux B, Hirt I (eds) The politics of mapping. Wiley, London, pp 91–126

Craib R (2009) Relocating cartography. Postcolon Stud 12(4):481–490

Del Castillo L (2018) Crafting a republic for the world: scientific, geographic, and historiographic inventions of Colombia. University of Nebraska Press, Nebraska

Donaldson JW (ed) (2008) Politics and scale in boundary-making: the work of boundary commissions (special issue). J Hist Geogr 34:393–515
Edney MH (1997) Mapping an empire: the geographical construction of British India, 1765–1843. University of Chicago Press, Chicago
Edney MH (2019) Cartography: the ideal and its history. University of Chicago Press, Chicago
Edney MH, Pedley M (eds) (2019) The history of cartography, volume four: cartography in the European enlightenment. The University of Chicago Press, Chicago and London
Erbig Jr J (2015) Imperial lines, indigenous lands: transforming territorialities of the Río de la Plata, 1680–1805. Doctoral Thesis, University of North Carolina, Chapel Hill
Erbig J Jr (2020) Where caciques and mapmakers met: border making in eighteenth-century South America. University of North Carolina Press, Chapel Hill
Escudé C (1989) Contenido nacionalista de la enseñanza de la geografía en la República Argentina, 1879–1986. In: Borón A, Faúndez J (eds) Malvinas hoy: herencia de un conflicto. Puntosur, Buenos Aires, pp 411–454
Ferreira M (2001) O Tratado de Madrid e o Brasil Meridional: Os trabalhos demarcadores das partidas do Sul e a sua produção cartográfica (1749–1761). Comissão Nacional para as Comemorações dos Descobrimentos Portugueses, Lisboa
Forêt P (2000) Mapping Chengde: the Qing landscape enterprise. University of Hawaii Press, Honolulu
Foucher M (1991) Fronts et frontières. Un tour de monde géopolitique. Fayard, Paris
Furtado JF (2017). Evolving ideas: J. B. d'Anville's maps of Southern Africa, 1725–1749. Imago Mundi 69(2):202–215. https://www.jstor.org/stable/48547831
García-Álvarez J (2013) Los mapas. In: Moreno J, Núñez XM (eds), Ser españoles. Imaginarios nacionalistas en el siglo XX. RBA, Barcelona, pp 315–363
García-Álvarez J, Puente Lozano P (2015) Las Comisiones Mixtas de Límites y las representaciones geográficas de la frontera hispano-portuguesa (1855–1906). Revista de Historiografía 23:67–100. https://erevistas.uc3m.es/index.php/REVHISTO/article/view/2756
García-Álvarez J, Puente-Lozano P (2017) Bridging central state and local communities' territorial visions: boundary commissions and the making of Iberian borders, 1750–1900. J Hist Geogr 57:52–61
García-Álvarez J, Puente-Lozano P (eds) (2022) Beneath the lines. Borders and boundary-making from the 18th to the 20th century. Springer, Cham
García-Álvarez J, Puente-Lozano P (2023) La construcción de las fronteras modernas: en el mundo hispánico: una perspectiva geohistórica. In: García-Hernán D, García-Álvarez J, Curnis M (eds) Fronteras del mundo hispánico: ideas, conflictos, identidades. Sílex Ediciones, Madrid, pp 109–127
García-Álvarez J, Puyo J-Y (2015) La aportación geográfica y cartográfica de las Comisiones de Límites luso-franco-españolas (siglos XVIII y XIX). Terra Brasilis (Nova Série) 6. https://doi.org/10.4000/terrabrasilis.1693
García-Álvarez J, Puyo J-Y (2019) Los primeros intentos de delimitación moderna de la frontera pirenaica: la Comisión franco-española Caro-Ornano y su legado cartográfico (1784–1792). Revista de Historiografía 30:15–44. https://doi.org/10.20318/revhisto.2019.4742
García-Hernán D, García-Álvarez J, Curnis M (eds) (2023) Fronteras del mundo hispánico: ideas, conflictos, identidades. Sílex Ediciones, Madrid
Gesteira H (2014) Um observatório astronômico nos confins da América Portuguesa-1750–1790. In Gesteira H et al. (eds) Formas do Império. Ciência, tecnologia e política em Portugal e no Brasil. Séculos XVI ao XIX. Paz e Terra, São Paulo
Goettlich K (2025) From frontiers to borders. How colonial technicians created modern territoriality. Cambridge University Press, Cambridge
Górny M (2022) Drawing fatherlands: geographers and borders in inter-war Europe. Brill—Schoningh, Padeborn
Gogate PP, Arunachalam B (1998) Area maps in Maratha cartography: a study of native maps of Western India. Imago Mundi 50(1):126–140
Gregory D (1994) Geographical imaginations. Blackwell, Oxford
Haller SC, Tapia AJ (2023) Hitos patagónicos: historia social de la Comisión Argentina de Límites con Chile (1881–1903). Revista Electrónica de Fuentes y Archivos 2(14):7–34. https://doi.org/10.70629/1853.4503.v2.n14.43948
Harley B (1988) Maps, knowledge and power. In: Cosgrove D, Daniels S (eds) The iconography of landscape. Cambridge University Press, Cambridge, pp 277–312
Herb G (1997) Under the map of Germany: nationalism and propaganda, 1918–1945. Routledge, New York
Herzog T (2015) Frontiers of possession: Spain and Portugal in Europe and the Americas. Harvard University Press, Cambridge
Herzog T (2017) Historical rights to land: how Latin American states made the past normative and what happened to history and historical education as a result. In: Carretero M et al (eds) Palgrave handbook on historical culture and education. Palgrave Macmillan, New York, pp 92–106
Infante MT (2013) Boundary disputes in Latin America. In: Peters A, Wolfrum R (eds) Max Planck encyclopedia of public international law. MPEPIL
Jöns H, Heffernan M, Meusburger P (2017) Mobilities of knowledge: an introduction. In: Jöns H, Meusburger P, Heffernan M (eds) Mobilities of knowledge. Springer, Cham, pp 1–19. https://doi.org/10.1007/978-3-319-44654-7_1
Kaya Z (2020) Mapping Kurdistan. Territory, self-determination and nationalism. Cambridge University Press, Cambridge

Leuenbeger C, Schnell I (2020) The politics of maps. Cartographic constructions of Israel/Palestine. University Press, Oxford

Lois C (2012) "La patria es una e indivisible". Terra Brasilis (Nova Série) 1. https://doi.org/10.4000/terrabrasilis.138

Lois C (2015) El mapa, los mapas. Propuesta metodológica para abordar la pluralidad y la inestabilidad de la imagen cartográfica. Geograficando 11(1)

Lois C (2019) De las fronteras coloniales del imperio hispánico en América a los límites internacionales entre Estados latinoamericanos independientes: génesis de la imposibilidad de un mapa político de Sudamérica consensuado. Revista De Historiografía 30:207–222

Lucena M (1993) Laboratorio tropical. La expedición de límites al Orinoco (1750–1767). Monte Ávila Eds-CSIC, Caracas

Manso C (2016) La cartografía de las comisiones de límites en América meridional. In: Moreno JM, Raffaelli S (coords) Dueños del mar, señores del mundo. Museo Naval, Madrid, pp 80–101

Melón MA et al (eds) (2019) Dinámica de las fronteras en período de conflictos: el Imperio español (1640–1815). Universidad de Extremadura, Cáceres

Newman D (2006) The lines that continue to separate us: borders in our 'borderless' world. Prog Hum Geogr 30(2):143–161

Novaes AR (2019) Maps in newspapers: approaches of study and practices in portraying war since 19th century. Brill, Leiden and Boston

Novaes AR (2025) Roasting the 'great divides': on the crossroads between map history and decolonial thinking. Imago Mundi 77(2):209–217. https://doi.org/10.1080/03085694.2025.2582328

Olsson G (1992) Lines of power/limits of language. University of Minnesota Press, Minneapolis

Parodi C (2002) The politics of South American boundaries. Praeger, Wesport

Penhos M (2005) Ver, conocer, dominar. Imágenes de Sudamérica a fines del Siglo XVIII. Siglo XXI, Buenos Aires

Pickles J (2004) A history of spaces: cartographic reason, mapping and the geo-coded world (1st ed.). Routledge, London. https://doi.org/10.4324/9780203351437

Prescott V, Triggs G (2008) International frontiers and boundaries. Law, politics and geography. Martinus Nijhoff, Leiden

Puente-Lozano P, García-Álvarez J (2022) Drawing the lines, knowing the borders: an overview of recent scholarship on boundary-making and modern regimes of territorialization. In: García-Álvarez J, Puente-Lozano P (eds) Beneath the lines. Borders and boundary-making from the 18th to the 20th century. Springer, Cham, pp 1–20. https://doi.org/10.1007/978-3-030-96904-2_1

Reguera AT (2007) La lucha postal por el territorio. Scripta Nova XI, 237. http://www.ub.es/geocrit/sn/sn-237.htm

Romero LA (ed) (2004) La Argentina en la escuela. La idea de nación en los textos escolares. Siglo XXI Editores, Buenos Aires

Raj K (2007) Relocating modern science: circulation and the construction of knowledge in South Asia and Europe. Palgrave New York, pp 1650–1900

Subrahmayam S (1997) Connected histories: notes towards a reconfiguration of early modern Eurasia. Mod Asian Stud 31(3):735–762

Seegel S (2012) Mapping Europe's borderlands. Russian cartography in the age of Empire. University of Chicago Press, Chicago

Standen NL (1994) Frontier crossings from North China to Liao, C. 900–1005. (Doctoral Thesis). University of Durham, Durham

Strandsberg J (2010) Territory, globalization and international relations: the cartographic reality of space. Palgrave MacMillan, London

Urteaga L (2007) Vigilia colonial: cartógrafos militares españoles en Marruecos (1882–1912). Bellaterra, Barcelona

Warmoes I et al (eds) (2016) Cartographie et Traités de Paix (XV-XXe siècle). Cartes et Géomatique 228:10–182

Williamson B (2023) Historical geographies of place naming: Colonial practices and beyond. Geogr Compass 17(5). https://doi.org/10.1111/gec3.12687

Woodward D (1975) Five centuries of map printing. University of Chicago Press, Chicago

Yee CDK (1994) Chinese maps in political culture. In: Harley JB, Woodward D (eds) The history of cartography, vol 2, Book 2, The University of Chicago Press, Chicago

Zárate CG (2020) La invención de la cartografía amazónica. Terra Brasilis (Nvoa Série) 14. https://doi.org/10.4000/terrabrasilis.7231

Zusman P (2025) Borders and territory. In: Halvorsen S (ed) Latin American geographies. An introduction. Routledge, London, pp 25–37

Part I
Negotiating and Fixing the Line

2 Cartographic Failure: The International Arbitration of the Franco-Brazilian Dispute (Bern, 1 December 1900)

Jean-Yves Puyo

Abstract

On 1 December 1900, the Swiss Federal Council issued its arbitration ruling concerning the border between Brazil and France on the Guiana Plateau, the subject of a two-hundred years dispute. The positions advocated by France were entirely rejected. This research examines the role of cartographic evidence submitted by both parties, which ultimately proved to be of little impact. Drawing on documentation held in French and Swiss diplomatic archives, which is largely unpublished to date, along with the extensive documentation produced by the two parties during this phase of arbitration, I trace the production, presentation and eventual rejection of cartographic materials to establish an international boundary.

Keywords

Franco-Brazilian dispute · French Guiana · Power of maps · *Uti possidetis* de facto · Treaty of Utrecht · Swiss Federal Council arbitration (Bern, December 1, 1900) · Paul Vidal de la Blache · Baron de Rio Branco

2.1 Introduction

On 5 December 1900, just four days after the publication of the Swiss arbitration ruling that dismissed France's entire territorial claims over the vast, disputed Franco-Brazilian territory in Guiana, the Marquis François de Ripert-Monclar, France's representative in Bern, sent a strongly worded letter regarding the outcome of the ruling to his foreign affairs minister, Théophile Delcassé. After reading the lengthy memorandum accompanying the award—over 830 pages, in German—he expressed his deep dismay at the arbitrator's bad faith. In Ripert-Monclar's view, France had been accused of a nearly 300-year period of pursuing, 'with an unwavering bad faith, the spoliation of the weak. While accepting that arbitration opened each party by submitting their case to arbitration, each of the parties undertakes to suffer possible condemnation', he argued that 'they are also entitled to a conscientious study of their

This work has been supported by the Spanish State Research Agency under Grant Number PID2020-114088GB-I00/AEI/10.13039/501100011033.

J.-Y. Puyo (✉)
Laboratoire TREE (UMR 6031 CNRS),
University of Pau and Pays de L'Adour, Pau, France
e-mail: jean-yves.puyo@univ-pau.fr

J. García-Álvarez and A. Reyes Novaes (eds.), *Mapping the Contested Line*, Historical Geography and Geosciences, https://doi.org/10.1007/978-3-032-29889-8_2

defence, and to respect for their good faith[1]'. It was sickening, he concluded, to witness such an 'an impassioned indictment, distorting the facts, omitting evidence that would exonerate the accused, or, if such evidence is mentioned, summarily dismissing it[2]'.

In the context of our research, it is not our intention to revisit the arguments developed by both sides on this occasion. First, several explanations for France's failure persist to this day: they range from the skill of Brazilian diplomacy in gaining the favour of the Swiss arbitrators (as suggested by the French press of the time) to the shortcomings in the scientific approach employed by the renowned French geographer Paul Vidal de la Blache. According to Guy Mercier, Vidal de la Blache's arguments met with failure because he had not presented a balanced case, addressing both the strengths and weaknesses of the French position. His scientific reasoning, Mercier observes, was undermined by Brazil's invocation of *uti possidetis* de facto, a principle that prioritizes possession based not on historic rights but on actual occupation—a concept we will explore in greater detail later (Mercier 2009). While we fully agree with this assessment, we clarify (for the first time, to our knowledge) where France's methodological weakness stemmed from: the result of unsound conclusions drawn from an earlier international arbitration regarding the border between French and Dutch Guiana (Russian arbitration, 1891).

Second, we identify the factors that explain this (undeniable) victory of *uti possidetis* de facto by examining the significance of cartographic documents in the evidence submitted by the two parties to the Swiss judges. As our analysis of the Swiss arbitration's concluding memorandum shows, these documents played only a marginal role in the final deliberation on account of their poor overall quality. Moreover, we would go so far as to suggest that the Brazilian side, unlike the French, was fully aware that these documents would not withstand scrutiny. By overwhelming the Swiss arbitrators with more than 220 cartographic documents, all of which were of questionable quality (and question them the French did), the Brazilians successfully undermined France's scientific argument based solely on cartographic evidence. Brazil instead favoured a historical and ethnographic study of the de facto occupation of the disputed territory.

2.2 A Persistent Lack of Geographical Knowledge of the Amazon River Basin

In the early seventeenth century, and much later than the Spanish and Portuguese, the English, Dutch and French attempted to establish themselves in South America between the Orinoco and the Amazon. This vast geographical area, which covers nearly 1.5 million square kilometres and includes the Guiana Plateau, represented at the time the most ill-defined frontier of the two Iberian empires. It was situated at the intersection of the equatorial line and the well-known latitudinal parallel that resulted from the Treaty of Tordesillas (1494). After 1604, following various short-lived French settlement attempts near what would become the city of Cayenne (founded in 1643), it was not until the expedition of Jean d'Estrées, Count of Estrées, in December 1676—which forcibly removed Dutch neighbours from the territory—that France's presence in Guiana was definitively established; it remains so to the present day.

Meanwhile, the Portuguese gradually consolidated control over the left bank of the Amazon by founding the city of Pará in 1616, a definitive rejection of Spanish claims to the area. They soon found themselves in conflict, initially with the Dutch, but mainly with the French, who were also present on the left bank. By the late eigh-

[1]Marquis de Ripert-Monclar, letter to Minister Delcassé, Bern, 5 December 1900; French Legation in Rio de Janeiro, series A, article 107, CADN. Translator's note: Our translation from the French. Unless otherwise stated, all translations of cited foreign language material in this article are our own.

[2]Ibid.

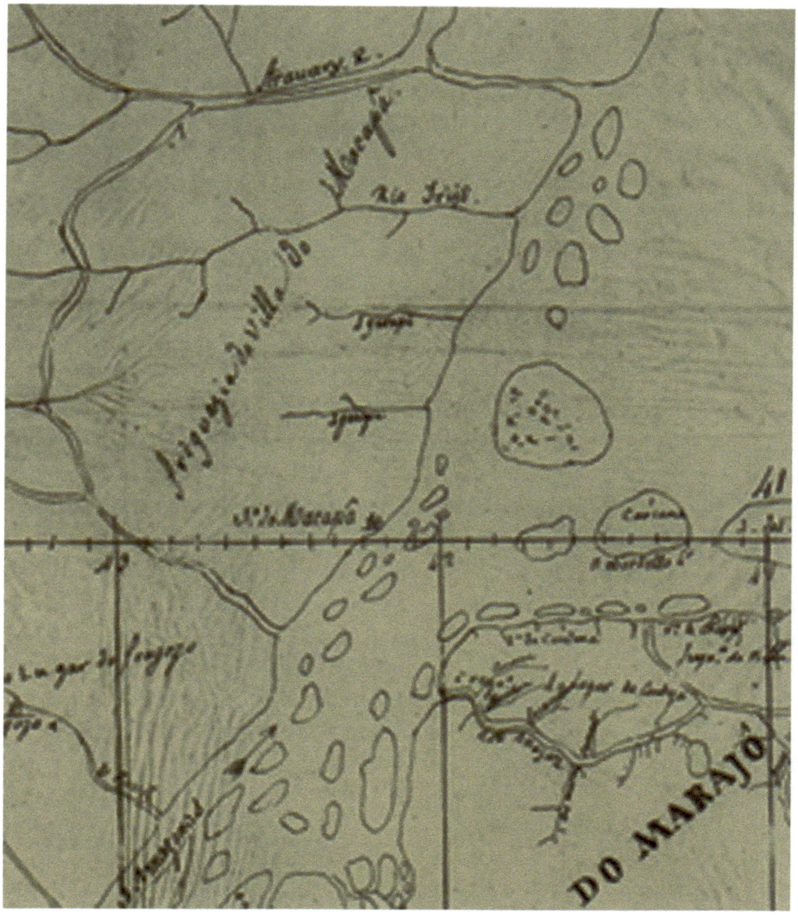

Fig. 2.1 Details of the mouth of the Amazon and the town of Macapá, extracted from a map of the northern part of the Bishopric of Pará (1759). *Source* French Republic (1899a, b, p. 75)

teenth century, the Portuguese settlement near the Araguary River,[3] including the establishment of Macapá (Fig. 2.1), was seen by the French as a territorial violation.

In response, in May 1697, a French military force captured Macapá and razed its fortifications before withdrawing a month later. A provisional treaty addressing the border demarcation between the two states was signed in Lisbon, on 4 March 1700. However, in the absence of an agreement between the two parties, the ownership of the lands of the 'North Cape' (Cap du Nord in French; Cabo Norte in Portuguese) was left unresolved 'until the conclusion of a definitive arrangement[4]'. Thus began the Franco-Portuguese dispute, known as the *Contesté franco-portugais* in French. The Treaty of Utrecht, which was signed on 11 April 1713, to end the War of the Spanish Succession, further complicated efforts to resolve this territorial issue.

The text of the treaty raised two major geographical issues. First, France was forced to withdraw from the mouth of the Amazon, which had been the primary objective for both Portugal

[3]There are several possible spellings for the name of this river depending on century and country. They include Araouary, Aroway, Arawary, Aroari, and Araguari, as well as Araguary, which was used in the nineteenth century. For the sake of simplicity, we will use the nineteenth-century form for place names.

[4]Baron de Damas, *Mémoire sur la contestation relative aux limites de la Guyane française*, addressed to the Comte de Gabriac, French ambassador to Brazil (report of 68 handwritten pages, dated 20 September 1826); French Legation in Rio de Janeiro, series A, article 102, CADN.

and England.[5] Although France now recognized Portuguese sovereignty over both banks of the Amazon (Article 10), the treaty failed to specify how far inland Portugal's sovereignty extended from the left bank.[6] France's territories, meanwhile, would begin north of a strip fifteen leagues deep, running parallel to the Amazon's course, and stretching east to west from the head of navigation of the Araguary River (which flows into the Atlantic off the North Cape) to the Rio Negro, a major tributary of the Amazon.[7] As the French minister of foreign affairs, Baron de Damas, noted in 1826: 'Apart from granting navigation rights on the Amazon, this treaty in no way altered the state of our former boundaries, which had previously extended as far as the Rio Negro, and thus the interior lands should continue to belong to France as far as this river[8]'. However, this position effectively meant controlling navigation on one of the Amazon's two major tributaries and, consequently, over a vast section of the Amazon basin upstream of Manaus. This territory thus defined corresponds to the Inland Contested Area (*Contesté intérieur*, Fig. 2.2).

Second, Article 8 ceded to Portugal 'ownership of the lands called the North Cape, situated between the river of the Amazons and that of Yapoc or Vincent Pinzon', but it failed to precisely locate this Yapoc (or Japock/Japoc) or Vincent Pinzon River (or Pinçon/Pinson River) named after the Spanish explorer Vicente Yáñez Pinzón: neither latitude nor longitude was provided.[9] According to the Portuguese and, later, the Brazilians, both names referred to the Oyapoc River, located 50 leagues north of the Amazon, between latitudes 4° and 5° north (Fig. 2.2). In contrast, the French argued that the river in question flowed into the 'Bay of Vincent Pinzon', located between latitudes 1° and 2° north, making it a northern branch of the Araguary that had later silted up, as described by the French naturalist Charles Marie de La Condamine after his visit to the area in February 1744.[10] This persistent disagreement gave rise to the Maritime Contested Area (*Contesté maritime*), which concerned both the nature of the much-disputed river and the so-called North Cape lands.

Indeed, as the Swiss judges responsible for arbitrating the territorial dispute later pointed out (an episode we will return to in detail), there were also fundamental differences of opinion between the two parties as to the identity of these same 'North Cape lands':

> France sees [in it] only the territory situated in the immediate vicinity of the North Cape; that is, the territory between the North Cape and the river of the Amazons, and wants to conclude from this that the "Japoc or Vincent Pinson" frontier river is situated in the south. Brazil claims that the designation "North Cape lands" in fact encompasses the entire region between the Amazon and Orinoco.[11]

[5]"At Utrecht, England was the mediator and almost the arbitrator between France and Portugal: from the dispatches and deeds drawn up at the time by her statesmen, it is clear that she wanted to ensure nothing but Portugal's exclusive ownership of the mouth of the Amazon and to close this river to the French" (Hamy 1901, p. 398).

[6]"His Most Christian Majesty recognizes by the present treaty that the two banks of the Amazon River, both the southern and the northern, belong in full property, domain, and sovereignty to His Portuguese Majesty; and promises [...] not to form any claim on the navigation and use of the said river, under any pretext whatsoever" (du Mont 1731, p. 35).

[7]"This line would follow the Amazon to the Rio Negro, then the left banks of the Rio Negro and Rio Branco, the Cakutu River and finally the last foothills of the Tumuc-Humac Mountains, which it would follow to the Maroni, leaving out the region currently disputed between Venezuela, England, and Brazil." Letter from the minister of colonies to the minister of foreign affairs, Paris, 28 April 1896; French Legation in Rio de Janeiro, series A, article 103, CADN.

[8]Baron de Damas, *Mémoire sur la contestation relative aux limites de la Guyane française*, op. cit., p. 15.

[9]Ibid.

[10]"A few leagues to the west of the Banc des Sept-Jours, and at the same latitude, I came across another mouth of the Arawi, now closed by sand. This mouth and the deep channel that leads to it from the north, between the mainland of the North Cape and the islands surrounding this cape, are the Vincent Pinçon River and Bay" (de La Condamine 2004, p. 126).

[11]Subsequently, the ruling was favourable to the Brazilian position: "There is nothing extraordinary in the fact that the word "Lands" has been used to indicate a fairly large

Fig. 2.2 Presentation of Brazilian and French claims. Extract from a map created for the Swiss arbitration commission's award (atlas comprising 5 maps and 3 tables, circa 1900). The border claimed by Brazil is shown in green, while France's claim is marked in red–orange. *Source* Swiss Federal Council (1900)

According to the Swiss arbitration commission, it was only after France had definitively abandoned the idea of a border based on the right bank of the Amazon in the Treaty of Utrecht that 'doubts gradually arose as to which river the said treaty had had in mind[12]'. Consequently, the decision under the Treaty of Utrecht was not implemented on the ground, and France continued its efforts to amend the terms it had conceded to Portugal.

2.3 "Post-treaty of Utrecht" and the Franco-Brazilian Border (1782–1895)

The question of determining the border between Portuguese and French possessions was of particular concern to the chancelleries of both countries at the nineteenth century. This period was indeed marked by a succession of treaties signed both at the local level between the governors of Cayenne and Pará (agreements of 1784 and 1792) and at the national level between the two central states.

Therefore, the geographical choices for establishing the border evolved according to geopolitical events and, consequently, the ebb and flow of their respective powers.[13] Later, the war with Portugal that began in November 1807 led to the occupation of what is now French Guiana by Portuguese forces from 1809 to 1817. The definitive resolution of the territorial dispute was further delayed by Article 107 of the Treaty of Vienna (8 June 1815), which stipulated that Portugal had to return the French colony up to the Oyapock River, a boundary that Portugal

and even a very large expanse of country" (Swiss Federal Council 1900, p. 749).

[12] Ibid., p. 775.

[13] The Treaty of Badajoz (6 June 1801) established the border demarcation on the Araguary River. The Treaty of Madrid (29 September 1801) introduced a separation along the Carapanatuba River, a tributary of the Amazon. The Treaty of Amiens (25 March 1802) reverted the border to the Araguary River, extended by a straight-line drawn westward, parallel to the Amazon's course, from its sources to the Rio Branco.

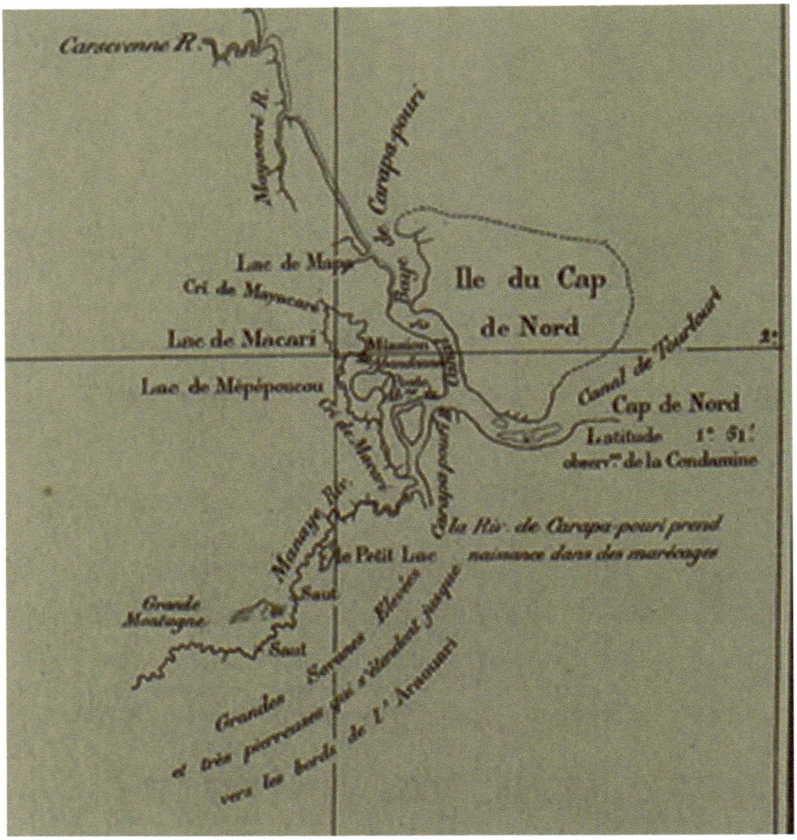

Fig. 2.3 Vicinity of the North Cape. Extract from the *Carte de la Guiane Française* (1788) produced by Simon Mentelle, from the atlas accompanying the French memorial submitted for the Swiss arbitration. *Source* French Republic (1899a, b, p. 89)

consistently considered as being defined by the Treaty of Utrecht. Whether Portugal was going to return the disputed territory depended on resolving the border issue. When the discussions failed, however, and Portugal was faced with the threat of a French military expedition, a new agreement was finally reached with the convention of 28 August 1817. This convention, which echoed the terms of the Congress of Vienna, set limits 'in accordance with the meaning of Article 8 of the Treaty of Utrecht[14]', which provided for English arbitration in the event of further disagreement—something France obviously wanted to steer clear of.[15] In practice, further disagreement was avoided: Portuguese troops evacuated the Guiana colony, and a long diplomatic truce followed, with Brazilian diplomats replacing their Portuguese counterparts in 1822, when Brazil gained independence.

In 1836, France took advantage of Brazilian political instability in the neighbouring federal state of Pará to push the limits of the August 1817 convention by establishing two military posts in the disputed area: one on the right bank of the Oyapock and another on the shores of Lake Mapa (Fig. 2.3).

Officially, these two settlements were intended to protect France's Guiana colony from a potential invasion by former independence rebels of Brazilian origin who had been expelled from Belém by Brazilian imperial troops. As the French minister of foreign affairs noted in correspondence with his representa-

[14] Baron de Damas, *Mémoire sur la contestation relative aux limites de la Guyane française*, op. cit., p. 23.

[15] "[English arbitration] would immediately arouse the jealous rivalry that this nation harbours against France, particularly with regard to French trade in America". Comte de la Ferronnays, minister of foreign affairs, Paris, 19 October 1836; series A, article 102, CADN.

tive in Brazil, however, the move was in fact designed to ensure France:

> [...] better chances of territorial gains beyond [what had been obtained under] the Treaty of Utrecht [...]. Let us truly occupy, with every appearance of permanence, the territories over which our rights of ownership are indisputable [...]. Then, when the time comes, we will be able to negotiate on the basis of the state of possession with even greater assurance, as public opinion will have been primed to accept our permanent retention of the territory we have occupied or colonized. (French Republic 1899a, b, p. 89)

This moment of weakness in Brazil's diplomacy was, however, short lived. After using military force to reintegrate Pará into its federal territory, Brazil demanded, in 1839, that the provisions of the Treaty of Vienna be applied to settle the dispute, once again threatening France with English arbitration. France therefore evacuated the two newly constructed forts between February and July 1840. By mutual agreement, the disputed territory was neutralized pending final delimitation, with France withdrawing to the left bank of the Oyapock. Shortly afterwards, Brazil took advantage of the situation to occupy the post of Mapa and create a new settlement, Pedro II, on the left bank of the Araguary—within the area claimed by the French, who at the time made no official protest.

In 1853, France sought to reopen discussions regarding the border demarcation. After months of delays caused by disagreements over the venue—France proposed Paris, while Brazil favoured Rio de Janeiro—the talks eventually resumed in Paris, in 1856, between the Viscount of Uruguai for Brazil and Baron His de Butenval for France.[16] Two years of intense discussions followed, during which the claims of both parties evolved significantly over time. Brazil abandoned its claim to the Oyapock and gradually shifted its position to rivers farther and farther north: first the Cassiporé River, then that of the Counani, and finally the Carsewène (Calçoene) River, whose mouth is located at approximately latitude 2° 35′ north (Fig. 2.4).

France, in turn, demanded a boundary on the Araguary, or, if its northern branch was found to be unnavigable, on the first river situated farther north. This alternative was identified as the Tartarugal, extended by the Mannaye and the Carapaporis rivers, with its mouth at latitude 1° 45′ north (Fig. 2.4).[17] As the international arbitration commission later noted, the two positions were separated by only 45 min of latitude, "therefore much less than one degree [...]" (Swiss Federal Council 1900, p. 460).

However, the negotiations ultimately ended in failure once again, as the parties did not resume talks "despite the wishes expressed by the two plenipotentiaries in the final protocol, and this promising attempt at conciliation remained without result" (Idem).

Subsequently, the vast Maritime Contested Area saw several noteworthy developments, including the short-lived Republic of Independent Guyana, also known as the Republic of Counani (1886–88), an experiment opposed by both France and Brazil (Fuligni 1997; Romani 2009; Puyo 2012). Six years later, the discovery of significant gold deposits along the Carsewène and Counani rivers (Fig. 2.4) triggered a massive influx of gold prospectors and other adventurers—estimated at 3000 to 4000 individuals[18]—into the disputed region. This led to the creation of several large settlements to serve as bases for gold panning activities (Romani 2013).

Tensions between French and Brazilian authorities escalated sharply as a result, culminating in a clash between French troops and

[16]France officially rejected the proposal to hold negotiations in Rio de Janeiro, citing a recent yellow fever epidemic that had claimed the lives of several foreign diplomats! However, selecting the French capital was also a strategy to avoid a setting that was perceived as advantageous to the Brazilian side. As Mr. Saint Georges, the French representative in Brazil, remarked at the time: "If this matter were dealt with in Brazil, I would not be surprised if it failed to produce the desired results." Letter to Drouyn de Lhuys, minister of foreign affairs, 14 July 1853, Rio; French Legation in Rio de Janeiro, series A, article 102, CADN.

[17]Letter from the minister of foreign affairs, 23 November 1856, to Mr. Saint Georges; French representative in Rio de Janeiro, series A, article 102, CADN.

[18]These included 'English' from English Guiana and the Caribbean; "Dutch" from Suriname; Brazilians; North Americans; and French from Guiana and mainland France.

Fig. 2.4 Maritime Contested Area, between the Counani and Araguary rivers. Extract from the map by Emmanuel Barralier, *Guyane française—d'après la carte publiée par la Société de géographie de Paris et la carte du Contesté par G. Brousseau*, 1900

"Brazilian irregulars" in the village of Mapa, on 30 May 1895. French forces suffered six fatalities, including the officer commanding the detachment, and twenty injuries, while Brazilian losses were estimated at 60. In retaliation, the French burned the village.[19] Faced with the risk of further military escalation, the two nations agreed, in August 1895, to establish a joint Franco-Brazilian commission 'to police and administer the disputed territories[20]'. Finally, in early 1897, the parties agreed to refer the matter to arbitration by the president of the Swiss Confederation. The signing of the arbitration agreement, on 10 April 1897, initiated the process of applying—with nearly two centuries of delay—the provisions of the Treaty of Utrecht, namely delimiting French and Brazilian possessions on the Guiana Plateau.

2.4 Borders for Young South-American States: *Uti Possidetis Juris* Versus *Uti Possidetis* De Facto

The early nineteenth century saw the emergence of new republics across South America following the dissolution of the Spanish

[19] "A Brazilian flag was flying from a flagpole in front of the village. The last order of the dying Captain Lunier [shot 5 times] was to take it away." Report of the confrontation published in the *Moniteur officiel de la Guyane française*, 22 June 1895; French Legation in Rio de Janeiro, series A, article 103, CADN.

[20] "Affaire du contesté franco-brésilien", anonymous note, Ministry of Foreign Affairs, Paris, 12 November 1895; French Legation in Rio de Janeiro, series A, article 105, CADN.

Empire.[21] At the Congress of Angostura, in 1819, these fledgling states agreed that their territorial boundaries would henceforth be based 'on the old borders and intra-imperial boundaries as they existed in 1810' (Sorel and Mehdi 1994, p. 14). This marked the adoption of *uti possidetis*, a legal principle holding that "the borders established under a now-defunct system must be respected and maintained by the new states" (Idem, p. 13).

Some authors have noted the paradoxical nature of this principle. On the one hand, the newly formed states could, in theory, claim nominal possession of territories over which they had no effective sovereignty at the time of their independence: 'They proclaimed a *fictitious imperium* over territories, arrogating to themselves the monopoly of future colonizations' (Barbagelata 1895, p. 4). For instance, Costa Rica and Colombia defended this position in 1899 during an international arbitration led by France: '*Uti possidetis* relates to territorial law, independent of occupation or possession' (Poincaré 1899, p. 3).

On the other hand, these same nascent states, formed through popular uprisings, could invoke the former colonial order to refuse the secession of provinces that wished to join other states. However, the principle was not universally applied by all of the newly independent nations. Mexico, Peru, Colombia and Argentina, for example, occasionally adhered to different rules for defining their borders, in accordance with the expressed wishes of local populations (Barbagelata 1895). There were also cases where boundary lines "set in stone" were blatantly disregarded, such as the numerous territorial reorganizations around the Rio de la Plata (today's Argentina) in the 1810s, which bore no relation to the boundaries inherited from the Spanish Empire.[22]

Nevertheless, by excluding European powers from the political division of the South-American continent, *uti possidetis* was to play a significant role in the development of international law concerning that same continent in the nineteenth century. On the other hand, it was rarely invoked in Europe. We find it in only two of the many major international treaties of the period: in the preliminary discussions for the Treaty of Paris (20 July 1806)[23] and in the Treaty of Erfurt (12 October 1808), between France and Russia.[24] We therefore hypothesize that French diplomats were relatively unfamiliar with the concept; this lack of familiarity was to have consequences later on, particularly during the Swiss arbitration of 1900. Brazilian diplomats, by contrast, worked throughout the second half of the nineteenth century to develop the principle of *uti possidetis*. At the time of its independence, in 1822, Brazil found itself in a somewhat paradoxical situation regarding its territorial limits, which were inherited from the eras of the Spanish and Portuguese empires. In the mid-eighteenth century, the two powers had reached a peaceful agreement to jointly define their South-American colonial possessions through a series of treaties: the Treaty of Madrid (13 January 1750), which was annulled by the Treaty of El Pardo (12 February 1761) and later replaced by the Treaty of San Ildefonso (1 October 1777). However, the war that broke out between Spain and Portugal in February 1801 disrupted this arrangement. The Treaty of Badajoz (6 June 1801), which ended the war, failed to restore the pre-war status quo, leaving Brazil and its neighbours without clear diplomatic boundaries at the time of independence (Accioly 1935). To invoke territorial rights in such a situation would be far-fetched; Brazil therefore modified the "old" principle of *uti possidetis* when it negotiated the

---

[21] Venezuela declared independence in 1806, Bolivia and Peru in 1809, Uruguay and Argentina in 1814, and Brazil in 1822.

[22] The province of Cuyo was established in November 1813 as part of the division of the former Córdoba government-intendency into multiple entities. These territorial reorganizations also saw the creation of the new provinces of Entre Ríos, Corrientes, and the Province Oriental del Río de la Plata in March 1814 (Barbagelata 1895, p. 6).

[23] England demanded that the French side negotiate "on the basis of *uti possidetis*, except for Hanover, which it insisted be returned"; and France refused (Ouroussow 1884, p. 176).

[24] Article IV: "the basis of the negotiations proposed to England will be *uti possidetis*" (Ibid., p. 189).

initial border treaties. For example, the Treaty of Ayacucho, signed in La Paz with Bolivia (27 March 1867), recognized *uti possidetis* as a rule of law in its second article. Later, during border demarcations carried out by an ad hoc joint commission, Brazil agreed to adjust the initial line to incorporate the former Bolivian colony of San Matías (covering nearly 6000 square kilometres north of Corixa Grande) into Bolivia's territory (Idem, p. 42).

This appears to be the first application of what would later be known as *uti possidetis* de facto, as distinct from "simple" *uti possidetis,* henceforth referred to as *uti possidetis juris.* In the case cited above, Brazil was not bound by *uti possidetis juris* because there was no "valid" agreement between the two former colonial powers, as previously noted: "Instead of *uti possidetis juris* of 1810, Brazil preferred *uti possidetis* de facto. According to this thesis, it was no longer a question of re-establishing an old boundary, but of reconstituting the de facto occupation of the territories as it existed in 1810" (Sorel and Mehdi 1994, p. 26). The nation subsequently applied this approach in all its border disputes with neighbours, including France in the Swiss arbitration.

2.5 Preparatory Framework for the Swiss Confederation's Deliberation and Lessons Learned from the Russian Arbitration of the Franco-Dutch Dispute (25 May 1891)

To properly evaluate the Swiss arbitration, it is essential to examine the nature of the 'mandate', namely the content of the arbitration treaty, which shaped both the arguments presented in the memorials of the two parties and the process leading to the ruling. In April 1897, after extensive negotiations, the two states agreed on the text of the arbitration convention, which comprises eight articles. Article 1 set out the claims of both sides regarding the interpretation of Article 8 of the Treaty of Utrecht: for Brazil, a northern border based on the course of the Oyapock River; for France, one based on the Araguary River. Article 2 outlined each party's expectations concerning the interior border: for Brazil, the border line proposed in 1817; for France, the one proposed in 1856. Notably, the convention allowed the arbitrator to define an intermediate line for these interior borders, which prompted the French ambassador to Brazil, Stephen Pichon, to write: 'This is an advantage for us[25]'.

Article 3 gave the two states eight months to submit their initial memorials once the arbitration phase had been officially initiated. Under Article 4, they would then have another eight months to submit their counter-memorials (the "2nd Memorial"). Thereafter, the arbitrator was allowed to request additional clarifications, if necessary (Article 5). Perhaps surprisingly, the treaty did not allow for direct debate in the presence of the judges, i.e. direct exchanges between the parties (Article 7). Finally, the arbitrator was given one year to deliver the ruling after receiving the counter-memorial (Article 8).[26]

After the formal ratification of the agreement on 6 August 1898, in Rio de Janeiro, the Swiss Federal Council accepted the arbitration mandate on 8 September 1898, at the request of both parties.[27] With the die now cast, the special-

[25]Stephen Pichon, letter to the Ministry of Foreign Affairs, Rio de Janeiro, 15 February 1897; French Legation in Rio de Janeiro, series A, article 103, CADN.

[26]Article 6 addresses the financing of the expenses incurred by the Swiss Confederation for the arbitration, with the total cost amounting to CHF 125,047.25, to be borne equally by the two states (Inventaire Commission Suisse, reference code 569/87519, 1087 & 1712, SFA). Notably, the content of the arbitration treaty was subject to debates before receiving official ratification by the parliamentary representatives of both states.

[27]The Swiss arbitration commission, chaired by Eduard Müller, president of the Swiss Confederation, included four principal experts, each of whom prepared an interim report. The synthesis of these reports—originally written in German and later translated into French—was the basis for the final report. The commission was made up of two geographers (Johann Jakob Früh from the Eidgenössische Technische Hochschule in Zurich and William Rosier from the University of Lausanne), an ethnologist (Otto Stoll from the University of Zurich), and a historian

ists from both countries began preparing their cases, though concerns soon surfaced on the French side. Shortly before the arbitration began, the French representative in Brazil expressed apprehension about the final outcome, noting that since the 1856 discussions, the Brazilians had been 'thoroughly familiar' with France's arguments and had studied them extensively, 'including their weak points', even publishing several works on the subject.[28]

France had just suffered a bitter failure in a similar arbitration, conducted by the Tsar of Russia, concerning the border between French Guiana and Dutch Guiana. All of France's arguments had been rejected, and the entirety of the Franco-Dutch Contested Area—the territory upstream of the Awa and Tapanahoni rivers—was awarded to the Netherlands. The French Ministry of Foreign Affairs sought to learn from this failure, but some of the lessons drawn were unsound, and these mistakes would have an impact on France's approach to the Swiss arbitration less than a decade later. First, by all appearances, ethnographic evidence relating to the actual possession of the contested area—which fell under the principle of *uti possidetis* de facto—did not have a decisive impact on the outcome of the Russian arbitration (Puyo 2024). In fact, although ethnographic evidence constituted roughly a third of the material submitted, it is clear that the French side ultimately attached very little importance to it.

Second, the content of the French brief was surprisingly weak: 'On reading the case for France's rights, one is in fact amazed at the general carelessness of the argumentation' (Bellardie 2006, p. 537)—just one map was included in the appendix, for instance. Following this failure, France was intent on preparing the memorials for the Swiss arbitration with greater care, and Paul Vidal de la Blache, the pre-eminent French academic geographer of the time, was recruited for the task.[29] For this phase of arbitration, the French commission for the contested area in Guiana produced three substantial memorials: two for the initial phase, accompanied by an atlas, and a third responding to Brazil's arguments from the first phase. The Brazilian side submitted three memorials for the first phase, accompanied by an atlas and a two-volume work published circa 1861–62 by Joaquim Caetano da Silva, titled *L'Oyapoc et l'Amazone: Question brésilienne et française.* For the second phase, in response to France's initial arguments, Brazil delivered no fewer than five volumes, and an accompanying atlas to boot.[30]

The cartographic collection assembled was substantial. Its purpose was to help meet the primary objective of the arbitration agreement: to determine the identity of the much-discussed watercourse referred to as the Japoc or Vincent Pinzon in the Treaty of Utrecht. The arbitration commission repeatedly emphasized this task in the detailed text of its award:

> Regarding the maritime boundary, the arbitration agreement necessitates an investigation into the precise interpretation of Article 8 of the Treaty of Utrecht. This requires an analysis of the treaty itself, supported by the scientific insights offered

(Albert Burckhardt-Finsler from the University of Basel). Additional support was occasionally provided by Paul Schweizer (professor of "historical auxiliary sciences" at the University of Zurich), Professors Woker (ethnologist at the University of Bern), von Salis (unidentified), an engineer named Schüle, and two translators, Charles Vogt and Charles Rousselet. Inventaire Commission Suisse, reference code 569/87519, 745, SFA.

[28]Stephen Pichon, letter to the Ministry of Foreign Affairs, Rio de Janeiro, circa October 1897; French Legation in Rio de Janeiro, series A, article 105, CADN.

[29]The French commission tasked with preparing the documents for submission to the Swiss arbitration included Desbuissons, a geographer from the Ministry of Foreign Affairs, his colleague Farges from the Historical Office, and Gabriel Marcel, assistant curator at the French National Library. The commission was later joined by Henri Félix de la Mothe, a former governor of French Guiana.

[30]Although José Maria da Silva Paranhos Júnior, the baron of Rio Branco and Brazil's representative at the Bern arbitration, signed off all the volumes produced, the hands that penned these documents are not known. However, the earlier writings of Joaquim Caetano da Silva played a significant role, as Carlo Romani highlights: "His efforts were not in vain. The Baron of Rio Branco relied on these two volumes to build his defence arguments in the legal dispute with France over possession of the disputed territory between the two countries" (Romani 2008, p. 50).

> by history and geography. [...] It is therefore necessary to examine [...] the maps that the parties, particularly the Portuguese, relied on during the Lisbon negotiations and to assess the information these maps convey. (Swiss Federal Council 1900, p. 18 and 796)

Based on this initial determination, the arbitrator would then be able to propose the delimitation of the internal boundary. Vidal de la Blache fully adhered to this approach:

> There is only one way to clarify this geographical question, and that is to examine the documents, particularly the maps, that may have served as a basis for this designation. It would even be sufficient, it seems, to consult the maps contemporaneous with the treaty, those that the negotiators may have referred to. (Vidal de la Blache 1902, p. 7)

To accuse Vidal de la Blache of conducting a purely scientific study, entirely detached from the economic and demographic realities of the Contested Area at the close of the nineteenth century, is to misunderstand the nature of his original mission (Granger 2011).

The two parties "inundated" the arbitration commission with reproductions of cartographic documents, particularly the Brazilian side, which submitted no fewer than three atlases (not including the maps included as illustrations in their memorials)—more than 180 documents in total. These ranged from the map of Juan de la Cosa (1500) to one created by the Brazilian captain Braga Calvacante (1896). Among the documents was the map by Alain Manesson Mallet (Fig. 2.5).[31]

For its part, France limited its cartographic submissions to a single atlas containing 35 maps (French Republic 1899a, b); these ranged from the Sebastian-Cabot map (1544) to that of Costa Azevedo (1863–64).[32]

2.6 Discussion

There are several points to be made regarding the role of maps in the final outcome of the Swiss arbitration. First, both states provided a significant number of cartographic materials, but Brazil submitted five times as many documents as France. This imbalance reflected a fundamental difference in their interpretation of the arbitration agreement:

> [For the French side] the counter-memorials should focus on addressing the arguments raised in the initial memorials [...]. This is further demonstrated by the fact that, once the second eight-month period expires, the written procedure is considered closed. While the judge may still request clarifications, the parties are no longer permitted to counter each other's arguments [...]. Introducing new theories for the first time in the second memorial, which cannot be reviewed or contested, would seem contrary to the spirit of the agreement. It is clearly a matter of moderation and good faith. (French Republic 1899b, p. 20)

France therefore chose not to introduce new evidence during the second phase of the arbitration, unlike Brazil, which submitted a voluminous new atlas that France was unable to respond to (Republic of the United States of Brazil 1899b). However, some inconsistency in the French position emerged when, around eight months after the close of the second regulatory deadline, it sent the judges a corrected map of its inland border claims. This move aroused the ire of Baron Rio Branco, who remarked: 'We could cause a scandal, but we will not, so confident are we that our rightful claim will be recognized[33]'. Finally, although the arbitration commission requested a handful of supplemental documents from both parties, these primarily consisted of older works absent from Swiss collections, not new maps.[34]

---

[31]For phase 1, Brazil submitted two atlases: the first included a collection of 91 maps dating from before the Treaty of Utrecht (see Appendix 1) (Republic of the United States of Brazil 1899a), while the second featured three maps produced by the Brazilian Commission for the Exploration of the Upper Araguary, led by Captain Braga Cavalcante (1: 200,000, 1896). And for phase 2, an atlas of 86 maps, including 14 prior to the Treaty of Utrecht (see Appendix 2) (Republic of the United States of Brazil, 1899b, c).

[32]See Appendix 3.

[33]Letter from Baron de Rio Branco dated 10 October 1899 to the arbitration commission; reference code 569/85719, 1097, SFA.

[34]For example, Gonzalo Fernández de Oviedo, *Historia general y natural de las Indias 1535–1548*, mentioned on page 256 of the first French memorial, requested from the French delegation on 13 November 1899 and subsequently forwarded on 22 December; reference code 569/85719, 701 & 712, SFA.

Fig. 2.5 Map of the *Pays des Caribes de Guiane* by Alain Manesson Mallet, 1683. *Source* Republic of the United States of Brazil (1899a, p. 303)

Second, the French commission relied almost exclusively on its national public resources for their cartographic material. These included, in order of importance, the collections housed in the National Library of France (BNF), the archives of the Ministry of Foreign Affairs, and those of the French Navy's hydrographic service.[35] In contrast, their Brazilian counterparts had to devote significantly greater efforts and financial resources to assembling their voluminous cartographic collection, sourcing the oldest maps from documents held in various European libraries and archives.[36]

Third, and unsurprisingly, the cartographic materials gathered by both states reflect the evolution of international cartography, which was dominated in the first half of the sixteenth century by the Spanish (Seville and Majorca) and Portuguese schools. This was followed by the emergence of Dutch, English, German and French cartographic traditions in later periods. Vidal de la Blache argued that 'as the origin of the name of the river Vincent is [to be found] in the official cartography of Seville, the truth about its position should be sought in the documents emanating from this source'

[35]For details concerning the history of cartography, Paul Vidal de la Blache cites his bibliographical sources, which primarily include the works of Henry Lavisse, Konrad Kretschmer, Gabriel Marcel, Adolf Erik Nordenskiöld, and Justin Winsor (Vidal de la Blache 1902, p. 10).

[36]Ibid.

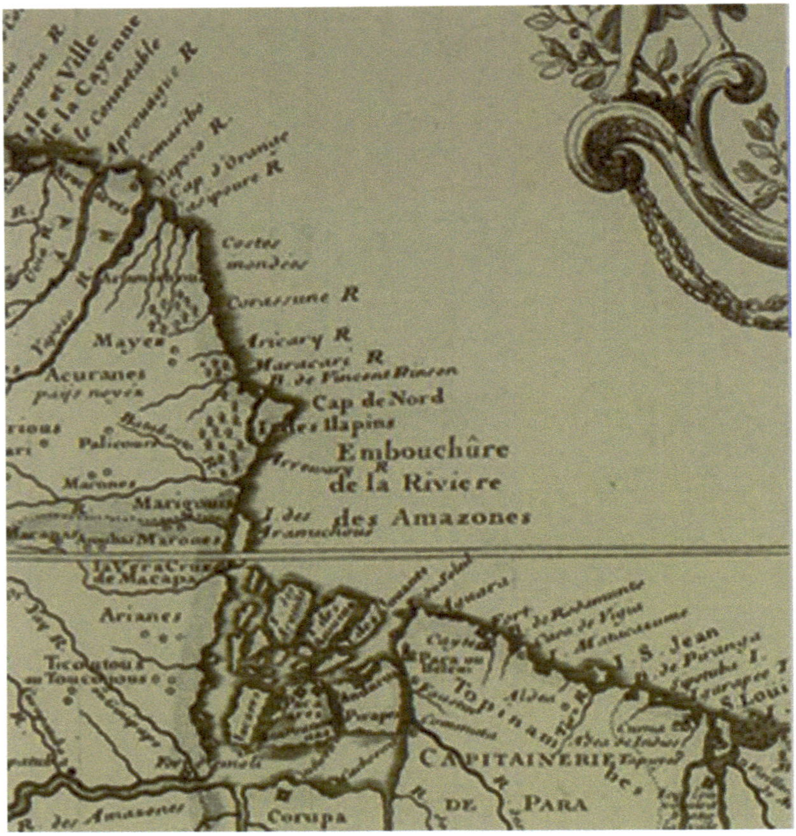

Fig. 2.6 Guiana Plateau, from "the Isle of Cayenne" to the mouth of the Amazon. Extract from Guillaume de l'Isle, *Carte de la Terre Ferme, du Pérou, du Brésil et du Pays des Amazones,* completed in 1703. *Source* French Republic (1899a, p. 51)

(Vidal de la Blache 1902, p. 47). For maps contemporary with the Treaty of Utrecht, on the other hand, Vidal de la Blache preferred maps from the late seventeenth century, particularly Guillaume de l'Isle's map, published in 1703 (Fig. 2.6):

> [It] reflects the French opinion on this matter during the same period. We cannot assume that the location assigned to Vincent Pinzon's Bay by Guillaume de l'Isle, on a map bearing his name and therefore carrying an almost official status, differed from the position attributed to it by the French negotiators. (Idem, p. 77)

It is worth noting that Brazil's response to these two points involved challenging the quality of both the Spanish cartographers' work and that of their French successors: 'The cosmographers of Seville in the sixteenth century should not be expected to have had more knowledge of America than the most eminent French cartographers of the seventeenth century had of their own country'—knowledge that was deemed sorely deficient.[37]

Nevertheless, the Swiss judges' assessment of the maps submitted by both parties revealed their considerable discomfort. William Rosier, for example, made no secret of his criticism after reviewing 'no less than 220 old maps': 'The atlases supplied by Brazil have not been composed methodically; in drawing them up, the aim was the number of maps rather than their scientific value; many, being mere copies, could have been left out' (Rosier 1900, p. 1097). Although France's submission included far fewer maps, it was likewise found to be methodologically lacking: 'Instead

[37] To demonstrate this, the Brazilian argument reproduced passages from a Belgian work discussing the difficulties encountered by Nicolas Sanson, Claude de l'Isle, and his son Guillaume in mapping the Mediterranean area (Republic of the United States of Brazil 1899b, p. 63).

of being arranged in chronological order, the maps should have been grouped by school, with the cartographic type of each series indicated' (Idem).

Rosier concluded from his study of the maps that the geographical arguments presented by both the French and the Brazilians led to a dead end: 'No map prior to the Treaty of Utrecht, nor any geographer, has sufficient authority to impose a definitive perspective' (Idem, p. 675). He dismissed, for example, the contradictory arguments regarding the supposed latitudinal location of the Vincent Pinzon River on the old maps, which tended to favour the French position: '[He emphasized] that at that time, people were generally mistaken about the latitude of this river, believing it to be closer to the Amazon than it actually is' (Idem, p. 731). Ultimately, Rosier set out to correct 'the error made by the negotiators of the Provisional and Utrecht Treaties regarding the position of the frontier river', convinced that the Brazilians had been well aware of this issue during the Paris Conferences of 1855–56 (Idem, p. 777). He argued that this awareness was why Brazil proposed 'a transactional solution, offering as a frontier first the watershed bounding the Oyapock basin to the east, then the Cachipour, the Counani, and finally the Carsavenne [...]. The intention of the contracting parties [of the Treaty of Utrecht] was not to place the frontier as far from the Amazon as the Oyapock appears on present-day maps' (Idem, p. 792).

Rosier also completely rejected the French demand for an internal border based on the drawing of an astronomical line '[...] more or less parallel to the Amazon, which would cut across some of the watercourses, direct or indirect tributaries of the left bank of the river' (Republic of the United States of Brazil 1899c, p. 27). According to Rosier, such a border would be unique in the world, '[leading] to an arbitrary and artificial division of the country that would contradict all geographical laws' (Rosier 1900, p. 799).

2.7 Conclusion

The verdict was handed down on 1 December 1900: The two delegations were sent a lengthy memorandum 'signed and sealed with the state seal' and written in German. This was accompanied by a six-page atlas and two tables of nomenclature. The outcome constituted an utter defeat of the French position (Fig. 2.7):

> I: "In accordance with the precise meaning of Article 8 of the Treaty of Utrecht, the river Japoc or Vincent Pinçon is the Oyapoc, which flows into the ocean immediately west of Cap d'Orange and whose thalweg forms the border line".
>
> 2: "From the main source of the Oyapoc River to the Dutch border, the watershed of the Amazon basin, which in this region is almost entirely formed by the ridge of the Tumuc-Humac Mountains, constitutes the inland boundary". (Swiss Federal Council 1900, p. 845)

Cartography played only a marginal role in the final outcome, as the award memorandum emphasizes: 'With respect to the detailed examination of the main maps, the arbitrator knows how cautiously old maps should be used as evidence' (Swiss Federal Council 1900, p. 462). Baron de Rio Branco shared a similar view: 'We cannot help saying once again that it is quite impossible to identify any position whatsoever if the only information available is the false geographical coordinates taken from the sixteenth and seventeenth centuries[38]'.

This perspective is somewhat surprising, given that Brazil submitted more than a hundred cartographic documents from the same period, including Manesson Mallet's 1683 map (Fig. 2.5). What could such a map contribute beyond an illustration of the limits of cartographic representation in that period?

For the arbitration commission, "the purely geographical questions raised by the identification of the Vincent Pinçon River with one of the watercourses of the Brazilian-Guyanese coastline could not be resolved solely with the help

[38] Baron de Rio Branco, letter to the president of the arbitration commission, 11 July 1900; reference code 569/85719, 912, SFA.

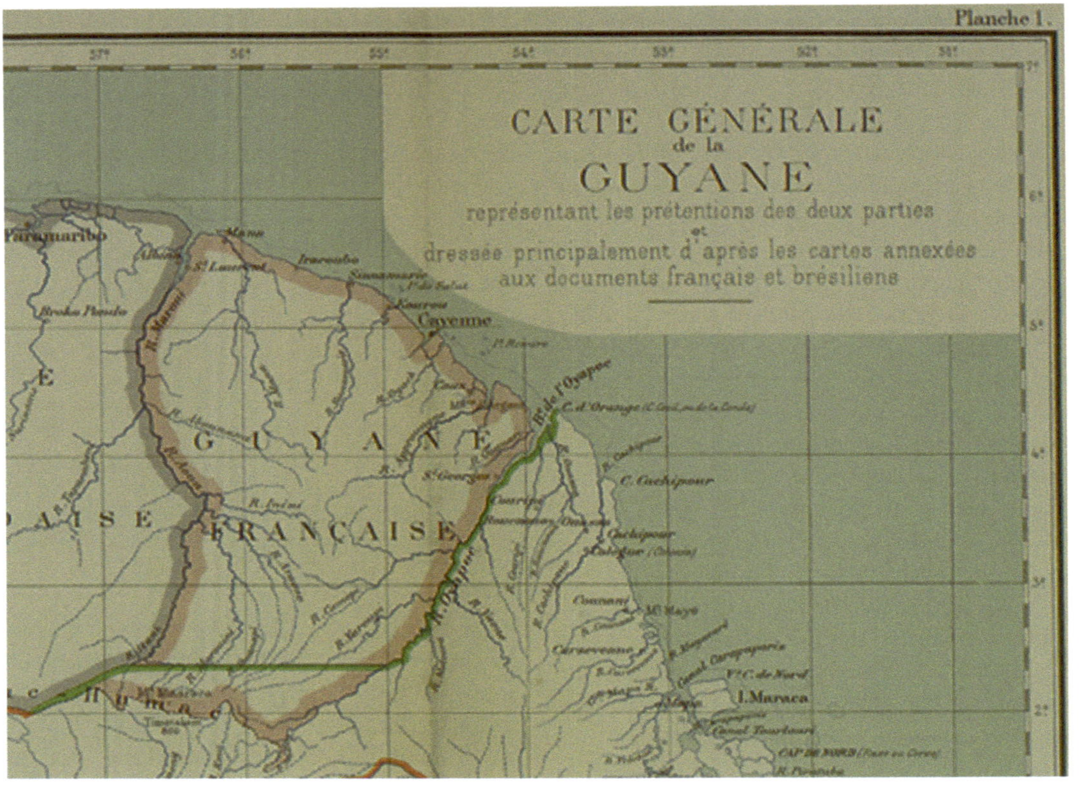

Fig. 2.7 French Guiana in its Swiss arbitration borders. Detail of the map created for the Swiss arbitration commission's award. The border requested by Brazil is shown in green, while the border established by the Swiss arbitration is marked in pink, a boundary that remains unchanged to this day. *Source* Swiss Federal Council (1900)

of maps dating from the time of the Treaty of Utrecht, so it was necessary to examine these questions in their relation to history [...]" (Swiss Federal Council 1900, p. 824). This led the commission to conclude that in Article 8 of the Treaty of Utrecht, the terms Japoc or Vincent Pinzon referred exclusively to 'the present-day Oyapoc at Cap d'Orange'—a categorical dismissal of the French position.

Two aspects of the decision warrant particular attention. First, William Rosier's proposal to 'correct' the supposed error in the 1713 treaty—an error that had resulted in the establishment of the maritime boundary at the mouth of the Oyapock—was rejected by his colleagues. Second, regarding the internal delineation of the new border, between the Maritime Contested Area and the Rio Branco valley, the arbitration commission noted that no decision had been made at the time of the Treaty of Utrecht. It therefore examined the rights of both states over this vast, sparsely populated, and largely unexplored region. Citing the works of French explorers Henri Coudreau and the renowned geographer Élisée Reclus, the commission concluded that Brazil had precedence over the territory, '[since it possesses] the lower course of all the rivers descending from it and all the points by which it can be reached from the Amazon. In accordance with the principle adopted by the powers for Africa, it forms part of the Brazilian hinterland' (Idem, p. 820). This argument likely came as no surprise to the French diplomats, who, in their correspondence just months before the start of the arbitration phase, expressed a fear of precisely this ruling:

> Expecting to secure an amicable settlement, whether through arbitration or any other means, for claims outside this zone [i.e. the Maritime Contested Area, situated between the Oyapock and the Araguary] would be entirely unrealistic. The entire Rio Branco valley, as Élisée Reclus observed, had become "indisputably Brazilian in language, customs, and political and commercial relations". As for the intermediate regions explored by Crevaux and Coudreau, they are inhabited by fully independent Indian populations.[39]

With respect to Coudreau's role in the final outcome of the Swiss arbitration, which Federico Ferretti argues was significant, we consider it to have been of secondary importance (Ferretti 2013). The Brazilians placed little value on the cartographic work of this French explorer.[40] This view was later echoed by the Swiss arbitration commission: 'From a purely cartographic point of view, Coudreau's indications do not always provide the requisite level of reliability' (Rosier 1900, p. 642).

French diplomacy gambled on scientific evidence, which it had lacked in the Franco-Dutch arbitration, by enlisting its most renowned geographer, Paul Vidal de la Blache—only to meet with a similar failure nine years later. France's scientific argument was defeated by Brazil's secret weapon: the invocation of *uti possidetis* de facto, a strategy that proved successful for Brazil in numerous border arbitrations in the same period.[41]

Archival Sources

CADN. Centre des Archives diplomatiques de Nantes.

SFA. Swiss Federal Archives (Bundesarchiv), Bern.

---

39 Stephen Pichon, French ambassador to Brazil, letter dated 23 August 1896 to the Ministry of Foreign Affairs; French Legation in Rio de Janeiro, series A, article 105, CADN.

40 "In addition to errors in latitude, there are discrepancies of 20 min in longitude between the coastline they show and that shown on the map, which is an accurate one [produced by Brazilian staff captain Braga Cavalcante in 1896]. In its response, France acknowledged that it lacked knowledge of the Upper Araguary and its tributaries (2nd French Memorial, p. 387)". Letter from Baron de Rio Branco dated 20 October 1899 to the arbitration commission; reference code 569 / 85719, 1102, SFA.

41 With Colombia (April 1907), Argentina (October 1908), Peru (September 1909), and Uruguay (October 1909). Brazil's single setback came in the arbitration ruling by the King of Italy regarding the dispute with British Guiana (Accioly 1935).

Appendix 1

Authors mobilized by the Brazilian atlas submitted to the Swiss Arbitration Commission at the first phase of the arbitration

Year	Author	Origin
1500	Juan de la Cosa	Spain
1519	Vescount de Maiollo	Genoa
1520	Ottomano Freducci	Italy
1523	Turin map	Italy
1527	Vescount de Maiollo	Genoa
1529	Diego Ribeiro	Portugal
1543	Nicolas Desliens	France
1544	Sébastien Cabot	England
1550	Diego Gutierrez	Spain
	Pierre Desceliers	France
1554	Jacopo Gastaldi	Venice
1558	Diogo Homem	Netherlands
1560	Spanish Map	
1561	Bertholomeu Velho	Portugal
1562	Bartolomeo Olives	Majorca
1563	Lazao Luis	Portugal
1564	Fernao Vas Dourado	Portugal
1568	Diogo Homem	Netherlands
	Fernao Vas Dourado	Portugal
1569	Gerard Mercator	Netherlands
1570	Abraham Ortelius	
1571	Fernao Vas Dourado	Portugal
1575	André Thevet	France
	Fr. de Belleforest	
1579	J. de Vaudeclaye	
1580	Fernao Vas Dourado	Portugal
	Bartolomeo Olives	Majorca
1582	Joan Martines	
1584	G. Battista Mazza	Italy
1585	Joan. a Doetechum	Netherlands
1587	Abraham Ortelius	
	Rumoldus Mercator	
1592	Théodore de Bry	Germany
1595	Michael Mercator	Netherlands
1596	Théodore de Bry	Germany
	M. Flor. van Langren	Netherlands
1597	Corlelis Wytfliet	
	Jodocus Hondius	

(continued)

(continued)

Year	Author	Origin
1598	Mathias Quad	Germany
	Barent Langenes	Netherlands
	English Map	
	Jodocus Hondius	Netherlands
1599	Levinus Hulsius	Germany
	Théodore de Bry	
	Petrus Plancius	Netherlands
	Ed. Wright, R. Hakluyt	England
1602	Gabriel Tatton	England
	Jodocus Hondius	Netherlands
1605	Cornelis Claeszon	
1606	Jodocus Hondius	
1608	Gabriel Tatton	England
1610	Harmen & Marten Janss	Netherlands
1614	Petrus Koerius	
1616		
1625	Johannes de Laet	
1628	Mathieu Merian	
1630	Ar. Flor. a Langren	
1631	G. Blaeuw	
1633	Jodocus Hondius	
1640	Clemendt de Jonghe	
	Joao Teixeira	Portugal
1642		
1646	Robert Dudley	England
1650	Nicolas Sanson	France
1655	Count of Pagan	
1656	Nicolas Sanson	
1657	Giuseppe Rosacio	Italy
1660	Dancker Danckerts	Netherlands
1664	Pierre du Val	France
1679	Guillaume Sanson	
	Pierre du Val	
1680	Guillaume Sanson	
	Claes J. Vooght	Netherlands
1683	Manesson Mallet	France
1688	Mr. Coronelli	Italy
1698	Froger & de Ferrolle	France
1700	Guillaume de l'Isle	
1705		
	Nicolas de Fer	
1707	Samuel Fritz	

Source Own elaboration based on Republic of the United States of Brazil (1899a)

Appendix 2

Authors mobilized by the Brazilian atlas submitted to the Swiss Arbitration Commission at the second phase of the arbitration

Year	Author	Origin	Observations
1536	Alonzo de Chaves	Spain	
1559	Andreas Homo	Netherlands	
1564	F. Vaz Dourado	Spain	
1570	F. Vaz Dourado	Spain	Manuscript atlas, British Museum
1576	Philippe Apian	Bavaria	Royal Library of Munich
1580	Bartolomo Olives	Spain	Royal Library of Spain (Madrid)
1588	Chr. Sgrothenus	Netherlands	Manuscript—Royal Library of Spain (Madrid)
1595	Glohe de Zurich	Switzerland	Mouths of the Amazon and Guyane (Zurich)
1618	Walter Ralegh	Netherlands	Manuscript map of Guyane—Archives of Simancas
	Lucas de Quiros	Spain	Manuscipt map—Royal Library of Spain
1625		Netherlands	Manuscript map—British Museum
1657	Giuseppe Rosacio	Italy	Engraved map—Naval Museum of Rotterdam
1658	Nicolas Sanson	France	
1660	Sebastian de Puesta	Spain	Manuscript map—British Museum
1714	C. Prize	Espagne	
1716	Samuel Fritz	Peru	Amazon River
1719	Anonymous	France	French map of the Amazon—manuscript
	Nicolas de Fer		«Part of his Terra Firma map»
	D'Anville		Map of French Guyane
1739	S.		«The Paris Skyline»
1741			Jesuit map of French Guiana es de la Guyane française
1742			Map of the French Navy Depot
1743	Pierre Barrère		Map of French Guiana
1745	La Condamine		Map of the course of the Maragnon or Amazon
			«His map compared to our modern maps»
1746	Le Rouge		Map of America
1747	João de Abreu Gorjão	Portugal	Lisbon Geographical Society
1748	D'Anville	France	South America
1750	Robert de Vaugondy		South America
1751		Portugal	Manuscipt map—Ministry of Foreign Affairs Madrid
			«Chart compared with our modern nautical charts»
1754	L. G. Desnos et J. B. Nolin	France	
1760	Jacques-Nicolas Bellin		Guiana—Maritime Atlas
1762	L. C. Buache		Depot of maps and plans of the Navy
	Jean Janvier		South America
	Bellin		Maps of the coasts of French Guiana
			map of French Guiana
			Navigation map from Cayenne to Martinique
1766	Philippe Buache		Handwritten sketch of a map of French Guiana
1771	Rigobert Bonne		«Mainland of Guyana and America»
1774	Bourgoin (d'Anville ?)		South America
1775	Cruz Cano y Olmedilla	Spain	Map of South America

(continued)

(continued)

Year	Author	Origin	Observations
1777	José Monteiro Salazar	Portugal	Handwritten nautical charts
			Lisbon Geographical Society
1780	Rigobert Bonne	France	French Guiana maps
			Northern part of Brazil
1781			New Andalusia and Guyana
			South America
	Biteow		French Guiana
1782	Jean Janvier		South America
			America
1785	Rigobert Bonne		«Guyane et du Pays des Amazones»
	W. A. Bachiene		New Granada, New Andalusia and Guyana
1790	Jean-Claude Dezauche		America
1791	Jean Baptiste Nolin		America
1797	Daniel Lescallier		French Guiana
1802	V. P. Malouet		French Guiana
1803	Mantelle & Chanlaire		French and Dutch colonies in Guiana
1808	Jean-Claude Dezauche		Map of America
1814	Jean-Baptiste Poirson	France	French Guiana
	P. Lapie		Map of South America
1815	A. de Beauchamp		Map of Brazil
	A. Lesage		«The two Americas»
1816	Adrien Hubert Brué		South America
	Brion de la Tour		
1820	P. Lapie		
1821	Adrien Hubert Brué		
1825	Vivien		
	J. A. Buchon		Brazil
	J. L. Darmet		South America
	Ch. Fr. Delamarche		America
	Jean Alexandre Buchon		French Guiana
1826	Aristide-Michel Perrot		French Guiana
	Adrien Hubert Brué		South America
			Brazil
	Louis Vivien		Map of Brazil
1828	P. Lapie		Colombia and Guyana
1829			Brazil
1834	Adrien Hubert Brué		South America
1838	P. Lapie		Map of Brazil
1839	Adrien Hubert Brué		South America
1846	C. J. de Niemeyer	Germany	
1849	V. Levasseur	France	
1853	Anonymous map	Brazil	
1860	Costa Azevedo		Hydrographic chart Disputed: Oyapock—Goyabal
			Hydrographic chart Disputed: Calçoene—Araguary
	French Nautical Chart	France	Cayenne—mouth of the Amazon

Note The presentation above responds to a desire to classify the cartographic documents according to a phasing divided into the following three periods: the maps prior to the Treaty of Utrecht; the productions of the eighteenth century; and, finally, the maps published since the end of the first French Empire

Source Own elaboration based on Republic of the United States of Brazil (1899b, c)

Appendix 3

Maps included the French atlas submitted to the Swiss Arbitration Commission

Year	Author	Origin	Observations
1544	Sébastien Cabot	England	Location of the Vincent Pinzon river
1550	Diégo Gutierrez	Spain	
1558	Diégo Homen	Portugal	The name Vincent Pinzon substituted for that of Rio-Fresco
1569	Gérard Mercator	Netherlands	Location of the Vincent Pinzon river
1595	Globe de Zürich	Switzerland	Mention of the Vincent Pinzon mouth (1° 30′ L N)
1598	A. Fl. Van Langren	Netherlands	Confusion Maragnon—Maragnan
1598	Richard Hakluyt	England	Mouth of the Vincent Pinzon at 1° 30′ L N
1618	Domingo Sanchez	Portugal	Vincent Pinzon's bay mention (1° 30′ L N)
1625	Jean Guérard	France	Yapoco at the North Cape
1627	João Teixera	Portugal	Northern limit of Brazil to the Para River
1630	J. de Laet	Netherlands	Guyanese cartography (English and Dutch travels)
1640	João Teixera	Portugal	Boundary of Portugal on the left bank of the Araguari
1647	Robert Dudley	England	Vincent Pinzon Bay near the North Cape
1656	S. d'Abbeville	France	French border at the Equator
1656	A. Roggeveen	Netherlands	Two mouths of the Araguari
1663	João Nunez Tinoco	Portugal	Vincent Pinzon River at about 1° 30′ L N
1689–91	Samuel Fritz	Peru	«Persistence of the oldest errors»
1695	Van Keulen	Netherlands	2 mouths for the Araguari
1703	G. de l'Isle	France	French border «a little north of the equator»
1717	Samuel Fritz	Peru	Copy of a map of the collection of *Lettres édifiantes*
1722	G. de l'Isle	France	«Boundary drawn in accordance with the Treaty of Utrecht»
1729	d'Anville		Vincent Pinzon Bay at North Cape—border at Araguary
1745	La Condamine		Vincent Pinzon bay on the northern mouth of the Araguary
	d'Anville		Northern border at North Cape
1749	Spain-Portugal Treaty		Manuscript –Vincent Pinzon bay at the mouth of the Araguary
1749–51			Engraved map: Vincent Pinzon River near Maraca Island
1750			Official map: French border at North Cape (1°–2° L N)
1753	Thomas Jefferys	England	French border at North Cape
1759	Êvéché du Para	Portugal	Northern boundary on the Araguary River
1763	John Gibson	England	Border on the Northe Cape
1770	Jan Jacob Hartsinck	Netherlands	Boundary on the Pinzon River (2° L N)
1783	Edw. Thomson	England	French Guiana's border on the Araguary River
1788	Simon Mentelle	France	French posts between North Cape and Orange Cape
1862–64	Costa Azevedo	Brazil	Topography and hydrography of the lands of the Cape

Source Own elaboration based on French Republic (1899a)

References

Accioly H (1935) Le Brésil et la doctrine de l'Uti possidetis. Revue Du Droit International 15:36–45

Bellardie T (2006) La Guyane oubliée: l'exemple du contesté franco-hollandais (1860–1891). In: Mam Lam Fouck S, Zonzon J (eds) L'histoire de la Guyane depuis les civilisations amérindiennes. Ibis Rouge Éditions, Cayenne, pp 523–540

Barbagelata HD (1895) Frontières: Contribution à l'étude de l'histoire du droit international américain. Librairie Paul Ollendorff, Paris

Ferretti F (2013) Le fonds Reclus-Perron et le contesté franco-brésilien de 1900: Une carte inédite qui a décidé des frontières du Brésil. Terra Brasilis (Nova Série). https://doi.org/10.4000/terrabrasilis.766

French Republic (1899a) Mémoire contenant l'exposé des droits de la France dans la question des frontières de la Guyane française et du Brésil soumise à l'arbitrage du gouvernement de la Confédération Suisse—Atlas. Phototypie Berthaud frères, Paris

French Republic (1899b) Réponse du Gouvernement de la République Française au Mémoire des États-Unis du Brésil sur la question de frontière soumise à l'Arbitrage du Gouvernement de la Confédération Suisse. Imprimerie nationale, Paris

Fuligni B (1997) Les constituants de l'Eldorado ou la République de Counani. Plein Chant, Bassac

Granger St (2011) Le Contesté franco-brésilien, enjeux et conséquences d'un conflit oublié entre la France et le Brésil. Les Cahiers d'Outre-Mers 372–373:157–177

Hamy E-Thr (1901) Oyapoc et Vincent-Pinson. Comptes rendus des séances de l'Académie des inscriptions et Belles-Lettres 1:396–402

(de) La Condamine Ch-M (2004) Voyage sur l'Amazone. La Découverte/Poche, Paris

Mercier G (2009) La géographie de Paul Vidal de la Blache face au litige guyanais: la science à l'épreuve de la justice. Annales De Géographie 667:294–317

(du) Mont J (1731) Corps universel diplomatique des droits des gens, vol. 8, contenant un recueil des traitez d'alliance, de paix, de trêve. Chez P. Brunel, Paris

Ouroussow A-M (1884) Résumé historique des principaux traités de paix conclus entre les puissances européennes, depuis le Traité de Westphalie jusqu'au Traité de Berlin. Charles Hérissey, Paris

Poincaré R (1899), Différend entre la Colombie et le Costa-Rica: Arbitrage de son Excellence M. le Président de la République française—deuxième mémoire présenté au nom de la République de Colombie. Motteroz, Paris

Puyo J-Y (2012) Du Contesté franco-brésilien à la République de Counani: histoire édifiante et curieuse d'une cryptarchie éphémère. In: Dornel L, Guicharnaud-Tollis, Parsons M, Puyo J-Y (eds) Ils ont fait les Amériques... Mobilités, territoires et imaginaires, Presses universitaires de Bordeaux, pp 279–291

Puyo J-Y (2024) La géographie, ça sert, aussi, à fixer des frontières: chronique d'un échec cuisant de la diplomatie française (le Contesté guyanais franco-hollandais, 1891). In: Buisson F., Durrans S., Caporossi O. (eds) Désirs, attraits et peurs des frontières: les Amériques dans tous leurs états, Presses Universitaires de Pau et des Pays de l'Adour, pp 59–74

Republic of the United States of Brazil (1899a) Frontières entre le Brésil et la Guyane française: Mémoire présenté par les États-Unis du Brésil au Gouvernement de la Confédération Suisse, arbitre choisi selon les stipulations du traité conclu à Rio-de-Janeiro le 10 avril 1897 entre le Brésil et la France, vol 1. A. Lahure, Paris

Republic of the United States of Brazil (1899b) Frontières entre le Brésil et la Guyane française—Second mémoire présenté par les États-Unis du Brésil au Gouvernement de la Confédération Suisse, arbitre choisi selon les stipulations du traité conclu à Rio-de-Janeiro le 10 avril 1897 entre le Brésil et la France, vol 2. Imprimerie Staempfli, Berne

Republic of the United States of Brazil (1899c) Frontières entre le Brésil et la Guyane française. Mémoire présenté par les États-Unis du Brésil au Gouvernement de la Confédération Suisse, arbitre choisi selon les stipulations du traité conclu à Rio-de-Janeiro le 10 avril 1897 entre le Brésil et la France, vol 3. Lahure, Paris

Romani C (2008) Algumas geografias sobre a fronteira franco-brasileira. Ateliê Geográfico (UFG) 2:43–64

Romani C (2009) Um eldorado fora de época: a exploração dos recursos naturais no Amapá. Trajetos: Revista de História da UFC 7(13):82–111

Romani C (2013) Aquí começa o Brasil! Histórias das gentes e dos poderes na fronteira do Oiapoque. Editora Multifoco, Rio de Janeiro

Rosier W (1900) Rapport et préavis présentés au Haut Conseil fédéral sur la question du Contesté de frontières franco-brésilien, Bundesarchiv (Bern), archive 569(85720):1097

Sorel J-M, Mehdi R (1994) L'uti possidetis entre la consécration juridique et la pratique: essai de réactualisation. Annuaire Français De Droit International 40:11–40

Swiss Federal Council (1900) Sentence du Conseil fédéral suisse dans la question des frontières de la Guyane française et du Brésil du 1er décembre 1900. Imprimerie Staempfli, Berne

Vidal de la Blache P (1902) La rivière Vincent Pinzon: Étude sur la cartographie de la Guyane. Félix Alcan, Paris

Fixing the Moving Line: Cartography, Territorial Issues and International River Boundaries—The Case of the Minho River Along the Spanish–Portuguese Border

3

Jacobo García-Álvarez

Abstract

The use of rivers to draw international borders between riparian states has been very frequent in contemporary times. However, such borders pose significant legal and geopolitical problems due to the fact that water is a fluid and dynamic phenomenon, while the concept of territorial sovereignty entails the ideals of fixity and permanence, as well as because of the difficulties of managing and administering this kind of transboundary areas in a unitary way. This contribution reflects on the role played by cartography in the processes of river boundary delimitation in general, and of conflicting river boundaries in particular, since the emergence and generalization of modern Boundary Treaties (18th–19th century). In order to do so, it first presents the main problems associated with the delimitation of international river boundaries, as well as the diversity of strategies that have been put into practice to fix these "moving lines" and the role that cartography has played or can play in these tasks. Second, as a case study, it examines the modern delimitation and demarcation of the Spanish–Portuguese border along the Minho river (1855–1898) and the main types of maps generated by this process.

Keywords

Cartography · Border disputes · Boundary delimitation · International river boundaries · Spanish–Portuguese boundary · 1864 boundary treaty · Minho river

3.1 Introduction

The use of rivers to draw international boundaries between riparian states has been very common in contemporary times. According to calculations based on the International River Boundaries Database (IRBD), correct as at 2007, nearly 76,000 km of the world's international land borders (equivalent to 30% of their total length), run along rivers and waterways of varying size and type, although this proportion differs according to the different regions of the planet. This represents approximately 43% of the total kilometres of land borders in South America, 32% of those in North America, 30% of those in Africa, 23% of those in Europe and 18% of those in Asia (Donaldson 2009,

This work has been supported by the Spanish State Research Agency under Grant Number PID2020-114088GB-I00/AEI/10.13039/501100011033.

J. García-Álvarez (✉)
Universidad Carlos III de Madrid, Getafe, Spain
e-mail: jacobo.garcia@uc3m.es

J. García-Álvarez and A. Reyes Novaes (eds.), *Mapping the Contested Line*, Historical Geography and Geosciences, https://doi.org/10.1007/978-3-032-29889-8_3

p. 641).[1] What's more, 57 of the 321 international land borders follow rivers for more than 70% of their length, and 20 boundaries are at least 99% riverine (Bishop et al. 2018, p. 12). Beyond specific cases, the general motivations behind this proliferation are as diverse as they are compelling. In brief, rivers make up clear and apparently simple geographical features to identify in the terrain and represent on maps; they can save considerable demarcation costs; they have served or can serve as defensive barriers in military or security terms; and they fit fairly well into the lineal form characteristic of the modern conception of border. In many colonial territories yet to be explored, rivers were, in fact, widely used as boundary markers because they were parts of the landscape that were more navigated, easier to access, and better-known than interfluvial regions.

At the same time, however, river boundaries pose major legal and geopolitical challenges and problems, which can be divided into two main categories. On the one hand, those associated with the planning, management and administration of the uses and resources of international rivers, and which may include, among others, issues related to infrastructure, navigation, fishing, energy exploitation, irrigation, tourism, flood control or pollution. And on the other hand, those stemming from the fluid and dynamic nature of water since, as Prescott and Triggs (2008, p. 215) have pointed out, '[t]he paradox inherent in conceiving of water flowing between two states as a territorial boundary is that the passage of the water is necessarily temporary, while sovereignty imports the notion of permanence'. As with other water environments, like lakes, seas or aquifers, rivers complicate our ideas of borders (Dodds 2021, p. 60). In the words of Donaldson (2009, p. 631), 'whether chosen for their defensive capability or their presumed clarity on the ground, river boundaries pose ongoing challenges for neighbouring states both in terms of identifying a definitive line in a dynamic natural feature and of managing a divided transboundary water resource.'

It is for these and other reasons (such as the fact that in many regions rivers have historically been axes of communication as well as of attraction and concentration of population and economic activity) that some of the main geographical contributions to boundary-making have strongly discouraged using waterways as border demarcations. For Stephen Jones, river boundaries hinder the unitary administration of the respective hydrographic basins and, at the same time, often raise the problem of identifying a stable border line, hence he recommended that 'river boundaries be avoided wherever possible.' (Jones 1945, p. 133). In more recent times, Biger (1988, p. 346) has considered that river boundaries 'are perhaps the most problematic of all boundaries, even though they might be seen as constituting a natural line, clearly marked on maps', while Gleditsch et al. (2006, p. 265) have stated that rivers are 'notoriously fickle' and 'fuzzy' boundaries, qualities that have led to and often increase the risk of interstate disputes, in some cases with military implications.

This chapter focuses on the role of cartography in territorial disputes associated with the delimitation of international river boundaries. To do this, it begins by addressing the dual challenge of: first, attempting to establish a lasting border in an intrinsically mobile space; and, second, in close relation to the foregoing, attempting to reconcile this type of boundary with the shared administration and management of the uses, functions and resources of such spaces situated between riparian countries. While it is true that, especially from the eighteenth century onwards, maps have frequently accompanied diplomatic negotiations and legal processes leading to the establishment of international borders, it is also true that the delineation of river boundaries poses a greater challenge for cartography: that of capturing in a static format, such as a map, an inherently dynamic phenomenon. Although it could be argued that all landscapes are subject to change, rivers undergo transformations that are particularly rapid, constant and visible, and there-

[1] The data broken down by interstate borders are available at: https://www.durham.ac.uk/research/institutes-and-centres/ibru-borders-research/maps-and-publications/databases/river-boundaries-database/.

fore the capacity of traditional maps to represent this dynamism is more limited. This difficulty is amplified when the aim is to map a river boundary where, as Biger (1988, p. 342) has stated, the tension between the physical (dynamic) geography of the river and the legal (static) dimension characteristic of the modern notion of border—what Donaldson (2011) has called 'the paradox of the moving boundary'—becomes especially evident.

The use of cartography in boundary delimitation, and within this in treaty and boundary conflicts, has been the subject of some general and comparative research (Blake 1995; Black 2009; Branch 2013; Warmoes et al. 2016). With respect to the delimitation of river boundaries, most of the general literature has come from international law (Bouchez 1963; Caflisch 1989; Schröter 1992; Prescott and Triggs 2008, pp. 215–232) with fewer contributions coming from the field of geography (Jones 1945; Donaldson 2011; Grainger and Conway 2014; Bishop et al. 2018). The vast majority of the studies in this regard have focused on the analysis of specific river boundaries, although few have paid attention to the cartography carried out in processes of delimitation (among the exceptions, see Carroll 2004; Donaldson 2010; Shelburne 2020; Costa 2017; Zusman and Nunes 2022; Duinat 2022). Finally, there are numerous studies focused on historical river cartography that, while not related directly to border delimitation processes, allow us to explore the ways in which maps have represented rivers and their complex dynamics over time (for example, Van den Brink 2000; Giménez 2009).

This chapter is divided into two parts, in addition to this introduction and the conclusion. In the first, general part, the main problems associated with river boundaries will be presented, together with the strategies and solutions used to try to establish these frontiers and the way in which they have been captured cartographically. In the second part, a case representative of the problems habitually associated with the delimitation of river boundaries will be examined: that of the Minho River, whose final stretch forms the dividing line between Spain and Portugal. Although this river has served as a border between the two countries since the independence of the kingdom of Portugal in the twelfth century, the modern delimitation of this zone did not take place legally until the Spanish–Portuguese Boundary Treaty of 1864 (henceforth 1864 BT), followed by an Annex in 1866 (on boundary rivers), and a Demarcation Act (or Act of Delivery), signed in 1897. Furthermore, the delimitation of the border line of the Minho River sparked numerous disputes and gave rise to a remarkably original, abundant and detailed body of cartographic materials, which illustrates the role that maps can play in the negotiation and resolution of border conflicts.

The analysis of this and of the territorial conflicts it gave rise to has already been the subject of previous research (García-Mañá 1993; García-Álvarez and Puente-Lozano 2015a, 2016, 2017). These studies have made it possible to identify the main disputes that arose in relation to the delimitation of the Minho border, the main arguments put forward by the Spanish and Portuguese governments (as well as by the local population), and the solutions provided by the 1864 BT and its subsequent developments. They have also demonstrated the role of the treaty as a flexible and context-specific tool for territorial management and have addressed some of the cartographic representations that it gave rise to.

Taking these works as its starting point, this chapter will focus on the role played by cartography to illustrate and interpret the problems associated with river boundaries; the strategies used to map these issues and, more generally, the changing landscape of the river; the production and use of maps to support the competing claims of the Spanish and Portuguese governments; and, finally, the contribution of cartography to the implementation of certain territorial planning measures provided for in the 1864 BT. To this end, the research carried out here will make use of the extensive archival documentation preserved in the Archivo Histórico Nacional de España (AHN) (National Historical Archive of Spain; located in Madrid), the Archivo Cartográfico y de Estudios Geograficos del Centro Geográfico del

Ejército (ACEGGCE) (Cartographic Archive and Geographic Studies of the Geographical Center of the Army; in Madrid) and the Arquivo Histórico-Diplomático do Ministerio dos Negócios Estrangeiros (AHDMNE) (Historical and Diplomatic Archives of the Ministry of Foreign Affairs; in Lisbon), including some previously unpublished cartographic documents.

3.2 International Rivers, Boundary Delimitation and Cartography

Biger (1988, pp. 341–344) and Bishop et al. (2018, pp. 16–33) have synthesized the main patterns of hydrology and the mechanics of rivers that can alter their course and affect the stability of international river boundaries. These changes can occur due to natural processes, human-induced processes (some constituting high-impact changes, such as the building of dams or the diversion of water for irrigation) or a combination of both. They arise from multiple factors (amount of flow and its variability and seasonality, valley slope, total sediment load, type of sediment, nature and amount of vegetation on the channel banks, human impact on these factors, etc.); and they are particularly notable in alluvial channels—that is, the lower stretches of the river that flow in a floodplain and whose bed and banks are mainly formed of alluvium (loose clay, silt, sand or gravel) that has been deposited by running water. In these stretches, the channel slopes are much gentler and the channel is liable to erode sideways into its floodplain sediments.

In summary, the two main changes that alluvial channels can undergo are accretion (gradual lateral movement of the channel due to sediment deposition) and avulsion (rapid lateral movement of the channel, which can be caused by massive erosive and sedimentary episodes linked to floods; the abandonment of a meander due to its high sinuosity; the presence of very small gradients in areas of high deposition, as typically occurs in deltas, etc.). In extreme cases of avulsion, sudden changes in runoff can shift the course of a river to create a completely new channel. In addition to lateral movements, a river can experience significant changes in the vertical plane (in which the river cuts deeper into its bed or the actual bed is filled with sediments), such as those that give rise to the emergence of river islands. Changes in a river's course occur at different time scales and, in some regions, they have a significant seasonal dimension. Some authors have also analysed the recent impact of climate change on some international rivers and lakes, as well as ice-covered boundaries, and the risk that this change may bring in the future on these and other aquatic boundaries (Grainger and Conway 2014).

States the world over have applied a variety of strategies and solutions to establish stable and permanent boundaries along international rivers, although two main types can be distinguished. The most important of these has been the establishment of commonly applied and widely accepted legal forms in international boundary law, which will be analysed later. The second most important strategy, sometimes used in conjunction with legal measures, has been to physically intervene in river courses prone to shifting via engineering works aimed at directing and stabilizing their flow—such as channelling, longitudinal dams, artificially induced cutoffs, alternative channels, as well as dredging and other operations to destroy natural and artificial obstacles.

Yet it is clear that most strategies and solutions aimed at addressing the problem of mobile river boundaries have focused on international law (previously the law of nations). The history of these solutions is long, since private Roman law already established mechanisms (*interdicta* or interdicts) to address the impact of changes in a river's course on the rights of riparian landowners (Sáry 2019). The distinction between the changes due to accretion (*alluvio* in Roman law) and avulsion (*avulsio*) was clearly established at that time. This distinction was later reflected in certain normative compilations of the Middle Ages (for example, the *Siete Partidas*, a Castilian statutory code written in the thirteenth century, but which remained in force in

Spain and Latin America right up to the nineteenth) and taken up by key authors in their landmark treatises on the law of nations of the seventeenth and eighteenth centuries, such as Grotius (*De jure belli ac pacis* [On the Law of War and Peace], 1625) and Emer de Vattel (*Le Droit des Gens* [The Law of Nations], 1758). These authors brought to the international arena a criterion of private Roman law that, without becoming a universally accepted norm (since there are numerous examples where it was not applied), was used in a number of modern boundary treaties, as well as in international arbitration. Specifically, changes in the courses of a river boundary caused by accretion must imply a modification of the boundary, such that it changes with the channel itself; while if the channel of a river boundary changes due to avulsion, the boundary remains along the 'original' bed of the river, even if this has since dried up (Donaldson 2011).

From the 1740s onward, rivers were widely used in boundary treaties to delineate territorial frontiers (Black 2009, p. 24). In parallel with this, the question of river boundaries took on increasing importance due to the increase in commercial navigation on rivers and, from the last third of the eighteenth century onwards, in light of the major changes to the European and American political map driven by certain historical developments, such as the expansion of revolutionary and Napoleonic France or the newfound independence of the majority of the American colonies. Thus, in the first decades of the nineteenth century, international boundary treaties and the main manuals of international law or law of nations already reflected a more or less consistent set of criteria for delimiting boundaries according to these natural landforms (Bouchez 1963; Caflisch 1989; Schröter 1992).

In summary, these criteria were, and have subsequently been, of four types, namely those that: (1) consider that the boundary is set on both banks of the river, and is therefore formed by the entire river, which itself belongs to border states jointly, as a kind of condominium; (2) define the boundary on one of the banks of the river so that all of it belongs to only one of the riparian states; (3) set the boundary on the median line, equidistant between the closest reference points or between the two banks and (4) define the boundary in the thalweg (a concept of German origin that has been defined variously as, among others, the main navigation channel, the median line or axis of the main navigation channel followed by sailors when descending the river, the centre of the deepest channel, the line along the deepest part of the river and where the current is fastest, or the line defined by the deepest points of the river bed). While the median line has often been used and recommended for boundary delimitation in non-navigable rivers, the thalweg has mostly been considered in the demarcation of navigable rivers—although both criteria have been used in both types of rivers, or have even been applied in the same river depending on the stretch considered. Finally, the delimitation of river islands has resulted in several legal solutions in international law (Bouchez 1963, pp. 815–817; Caflisch 1989, pp. 75–80; Schröter 1992, pp. 972–978).

In any case, the use of the four criteria indicated above has by no means been problem-free. The first issue stems from the vagueness, ambiguity and heterogeneity surrounding the use of some of these terms in diplomatic texts (the term 'thalweg' is a case in point, but also the use of the terms 'bank' and 'main channel', which in many cases has generated discrepancies and the subsequent need for legal clarifications). Other recurrent difficulties have resulted from the inherent complexity of hydrology and fluvial geomorphology. The bank of a river varies in height depending on the volume of water and can recede with erosion, while the median line also shifts depending on the water level. As for the thalweg, as we saw in the above definitions, this term has been used in some treaties as an area (navigation channel) and in others as a line (defined by the points of greatest depth or by the median or axis of the main navigation channel). Furthermore, the determination of the thalweg is not straightforward in rivers that branch into two or more channels due to the presence of intermediate islands, and in some rivers the main navigation channel does not coincide with the deepest part of the riverbed. Also, if the territorial jurisdiction over river islands depends on the position of the median or the chosen thalweg, any variation of that position could, in certain cases, lead to a redefinition of the sovereignty of those islands. Similarly, these islands can emerge, grow, shrink or disappear depending on channel variations. It is precisely owing to these and other complications that most boundary treaties contain rules that expressly prohibit states from modifying the course of the river via artificial means without the consent of both parties.

Finally, the use of these criteria alone does not solve the problem that the position of riverbanks, the median line or the thalweg can shift at a varying rate according to changing fluvial dynamics. Although certain treaties and arbitration decisions have followed the criterion mentioned above of differentiating the changes produced by accretion from those caused by avulsion, others have opted for specific solutions to address this issue, such as periodically reviewing the position of the thalweg in order to adjust the boundary line if any change has occurred; or establishing a fixed and definitive border based on the position of the river's course at a given moment, regardless of any change that this course may undergo afterwards. With respect to this latter option, some river boundaries have been delineated by means of invariable astronomical or geometric limits, drawn from a series of coordinate points (Bouchez, 1963, pp. 799 et seq.; Schröter 1992, pp. 964–972). This methodology, facilitated by GPS technology, has been followed in some arbitration decisions handed down by the International Court of Justice (Donaldson 2011, p. 167). Ultimately, to a large extent, the solution adopted in river delimitations depends not only (or less so) on technical criteria but also (or above all) on the geopolitical interests and priorities of the states involved, as well as on the power relations between them.

Although it began earlier, the use of maps in border delimitation processes became frequent among the main European powers from the middle of the eighteenth century onwards, both in metropolitan and colonial territories (Black 2009). In this respect, cartography already served various purposes at that time: to bolster the interests and territorial claims of the different parties involved, as supporting evidence; to provide useful information about the affected border areas, especially those subject to litigation and negotiation; and to represent as precisely as possible the border finally agreed upon, delimited, and demarcated, often accompanying an international treaty and thereby acquiring legal value. In addition, given the need to contribute to the delimitation of river boundaries, the maps used in these processes had to incorporate certain techniques specific to hydrographic cartography, as well as perfect or develop other techniques to meet the challenge of accurately depicting the unique issues associated with this type of space.

Some authors have also focused on the relationship between the technical progress of cartography, the increasing use of maps in diplomacy and the consolidation of a territorial concept of sovereignty, based on states delimited by fixed linear borders that enclose contiguous and compact territories and clearly divide the areas of exclusive sovereignty between these states (Biggs 1999; Branch 2013, pp. 120–141). In the opinion of Jordan Branch (2013, pp. 135–138), the Congress of Vienna and the Treaties of Paris of 1814–1815 ushered in the triumph of a linear conception of borders in the European context, as well as the systematic use of maps to enforce and make tangible these frontiers in the physical landscape. The liberal revolutions that swept through Europe in the nineteenth century further consolidated this trend, as they drew on a territorial notion of sovereignty and the nation, understood as the political association of citizens within a territory. The 1864 BT is emblematic of the transfer of these conceptions to the Iberian context, as well as the intense and complex relationships between river boundaries, territorial conflicts and cartography.

3.3 Maps and Territorial Issues Along the Minho River Border

3.3.1 The Spanish–Portuguese Border Along the Minho River: Historico-Geographical Context

The section of the Spanish–Portuguese border traced along the Minho River runs for approximately 75.5 kms from its mouth until it meets the Barjas or Troncoso River, a tributary to the south. This fluvial stretch accounts for just over 6% of the total length of the Spanish–Portuguese border (1232 km), of which, according to the IRBD, close to 55% (667 km) is defined by watercourses, such as the Douro (112 km), the Tagus (47 km) and the Guadiana (100 km), as well as the Minho. The Minho border, like the Spanish–Portuguese border in general, is of great antiquity. This river formed the northern limit of the *Condado Portucalense* (County of Portugal), created at the end of the eleventh century, as well as of the Kingdom of Portugal following its independence from the Kingdom of León, which began de facto in 1129 and was confirmed de jure in 1179. Since then, the Minho has remained a political border, with the exception of certain periods of war and the union of the Portuguese and Spanish crowns between 1580 and 1640 (Iberian Union), during which the political role of the fluvial boundary was suspended or weakened in many respects. In any case, and as already pointed out, the modern delimitation of this zone was not carried out until the Spanish–Portuguese Boundary Treaty signed in Lisbon on 29 September 1864 (United Nations 1991, pp. 275–283). This treaty traced the border line from the mouth of the Minho River to the junction of the river Caya with the Guadiana, and it was supplemented by two annexes signed in 1866 (relating, respectively, to the boundary rivers between the two nations and the seizure of livestock). The demarcation of the border defined in the 1864 BT was not completed until 1906, with the signing of the General Act of Demarcation, although the Act of delivery for the section delimited by the Minho River was signed in 1897.

The Minho is a fast-flowing river with a pluvial-oceanic regime, experiencing dry summers and rainy winters that can sometimes lead to flooding. Its international section runs in a general NE-SW direction through a mostly narrow and deep valley, with widths that, with important seasonal variations, range from a minimum of 20 m to a maximum close to 2 km shortly before its mouth in the Atlantic Ocean. In the middle of the nineteenth century, the river was navigable up to about the Portuguese town of Monçao, 44 km upstream from the mouth.

The technical preparation work for the 1864 Boundary Treaty was entrusted to a Joint Boundary Commission (henceforth JBC) formed initially by diplomats and military engineers from both kingdoms (later substituted by officers of the Army General Staff) who travelled along the border between 1855 and 1863. The stretch of the border formed by the Minho River was the first to be recognized by the committee, although it had to be reviewed several times (either jointly or separately along the Spanish and Portuguese sections) due to the disputes it caused. These disputes centred around three main issues. First, determining the sovereignty over certain river islands, whose ownership was the cause of conflict and confusion. Second, establishing a general criterion for tracing the dividing line defined by the river (and by extension, for all the other rivers along the border between the two kingdoms). And finally, resolving the problem caused by the proliferation of fisheries on both banks of the river. In the next section, the characteristics of these three points of contention, as well as some of the cartographic representations they gave rise to, will be expounded in turn.

3.3.2 Disputed Sovereignty Over Certain River Islands

With regard to the question of the Minho islands, the main territorial conflicts involved, on the one hand, Canosa island, located near the mouth of the river, opposite the town of Caminha, about which there seemed to be a long-standing agreement regarding its shared use between the Spanish and Portuguese; and on the other hand, the islet of Fillaboa ('Filha Boa' in Portuguese) and the Verdoejo group of islands, located in the stretch between Tui (Spain)/Valença (Portugal) and Salvaterra de Miño (Spain)/Monçao (Portugal). Within this group, the disputes primarily concerned the islands Cancela (also called 'Caldelas' and 'Cancella'), Ínsula Grande (also called 'Insua Grande' or 'Isla Grande'), Canguedo and Raña Gallega. The Spanish and Portuguese historically disputed the use and sovereignty of these islands, most of which were created or at least significantly altered due to the uncontrolled construction of numerous artificial works on both banks of the river, such as mills, small dams, embankments, jetties and, above all, fisheries made with wooden stakes or thick stone walls arranged perpendicular to the river, mainly for salmon and lamprey.

The issue of sovereignty over these river islands, as well as their origin, was hotly debated during the negotiation process of the 1864 treaty, leading to major disagreements, not only between the delegations of both countries but also within the Portuguese section itself (García-Álvarez and Puente-Lozano 2016). In turn, these disagreements generated a wealth of historical, geographical and cartographic documentation, much of it aimed at justifying the respective interests of each party. The map drawn in September and October, 1855, by the military engineer Ramón Madina-Orbeta, member of the Spanish section of the JBC, and the one produced in February 1859 by major José Maria Delorme-Colaço, secretary of the Portuguese section of the JBC, are prime examples of this.

The first of these maps (Fig. 3.1), accompanied by a detailed descriptive report written by Madina himself and collected by the diplomat Fidencio Bourman, president of the Spanish section of the JBC, gained particular importance in the subsequent negotiation process (AHN, doc. 1). It became a key tool not only for visualizing the main issues that this border stretch presented, but also for substantiating some of the arguments put forward by the Spanish delegation and proposing a series of solutions that were ultimately accepted in the 1864 BT regarding the disputed islands. Among other information, Madina's map depicted several of these islands (such as Cancela and the Verdoejo group), along with two topographical profiles of the river where these islands are located, and several constructions located on the banks—most notably, fisheries on both sides and the salt warehouses used by the Portuguese.

Based on the map and the topographical profiles mentioned, as well as on documents and testimonies collected on the ground, the Spanish engineer analysed the human and natural factors contributing to the formation of the islands and the potential problems they posed in terms of sovereignty, security or navigability. With regard to Ínsula Grande, for example, which is located very close to the right bank of the Minho, Madina argued that the de facto dominion of the Portuguese over the island gave them control of the river's waters and flow in that stretch, a situation that was not only detrimental to Spanish interests but also posed a military risk: 'it is such', Madina warned, 'that if it suited their future plans and interests to build a military stronghold or fortress, [the Portuguese] would have the immense advantage of commanding both banks of the river, as can be seen in some fortifications in Germany situated along the Rhine.' Likewise,[2] the Spanish commissioner concluded that, in recent years, the river had significantly changed its course, diverted by the fisheries and jetties built on the left bank by the Portuguese, causing flooding and erosion along the Spanish bank, while the Portuguese bank had gained ground. Therefore, the natural and ancient channel (which Madina called the 'true channel') was the one that ran between the island and the Portuguese bank, and it was there that the border between the two countries should be established, meaning that this island had to be

[2] Author's translation from the Spanish. Unless otherwise stated, all translations of cited foreign language material in this chapter are the author's own.

Fig. 3.1 *Plano de un trozo del río Miño desde la plaza de Tuy a Caldelas...* [Map of a section of the Minho River from the military stronghold in Tui to Caldelas], by Captain of Engineers Ramón Madina-Orbeta, dated in Ourense on 1 October 1855. Scale of 4000 feet for the map and 400 feet for the farmhouses. Detail of the stretch corresponding to the Verdoejo islands, north of the Portuguese village of the same name (Berdoexo on the map) and south of the Spanish parish of Santiago de Baldráns. The letters indicated on the map represent: A: Ínsula Grande; B: Canguedo; C: Raña Gallega; a: Portuguese salt warehouses; b: isolated houses; c: fisheries; and y: portions of land reclaimed from the river, on the Portuguese side, with the construction of fisheries. *Source* AHN, Ministry of Foreign Affairs Section, Maps and Charts, 1138

'returned' to Spain by the Portuguese. Furthermore, Madina attributed the greater number of fisheries on the left bank of the river to 'the substantial profits [the Portuguese] make from smuggling, especially of salt', activities that required 'leaving islands in the middle of the river and creating bends everywhere in order to evade the watchful eyes of the Carabineros [Spanish border guards]'.

In view of the map and Madina's report, as well as the report by Fidencio Bourman, the Spanish government instructed Bourman to claim Cancela and Ínsula Grande for Spain, given that their clear proximity to the Spanish bank (AHN, doc. 2). This proximity, these instructions insisted, should be the main argument used for claiming the islands, even over and above the documentary approach of clarifying the legal basis of ownership to determine whether or not they fall under Spain's dominion. Since the border that separates the jurisdiction of both states must run through the 'centre of the waters', and this main channel lies between the two islands and the Portuguese bank, these would fall under Spanish rule and therefore be part of its territory.

This criterion was initially accepted in June 1856 by the Portuguese commissioners;

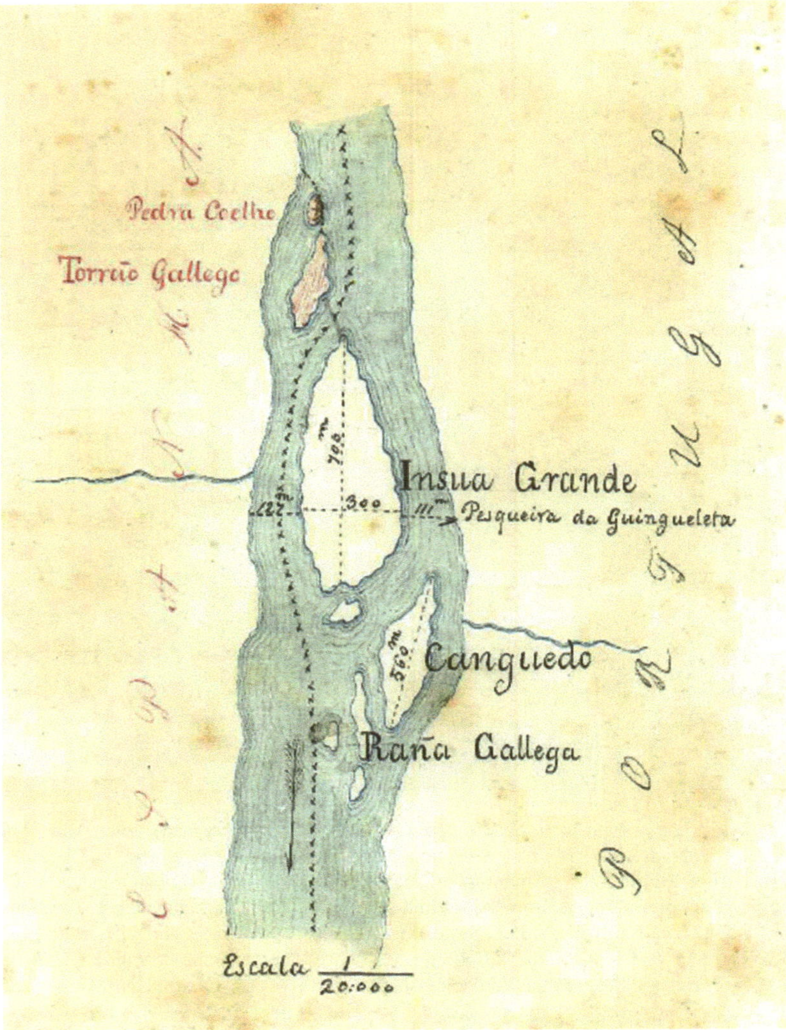

Fig. 3.2 *Planta das Ilhas do Verdoejo, no Rio Minho, levantada no mês de Fevereiro de* 1859 [Map of the Verdoejo Islands, on the Minho River, produced in February 1859]. Scale 1:20,000. *Source* AHDMNE, Boundary Commission Letters, Sig. 26-A. Unsigned map based on the one produced in black and white by José Maria Delorme Colaço (AHDMNE, Box 1119, Fol. 232 and 317)

however, three years later, a report prepared by Delorme-Colaço and the engineer José de Vasconcelos e Sá (president of the Portuguese section of the JBC since July 1858), along with the map Delorme-Colaço produced concerning the Verdoejo islands, challenged and proposed a review of this decision.[3] Delorme-Colaço and Vasconcelos contended, on the grounds of various historical and geomorphological arguments, that Ínsula Grande had originally been a single island, under Portuguese sovereignty, which had apparently broken apart as a result of works built on the Spanish bank, giving rise to the islands Canguedo and Raña Gallega. For both commissioners, the border between both kingdoms should not be set along the current thalweg—created by artificial works built on the Spanish bank—but along the 'natural' or ancient course of the river prior to this alteration.

In support of this thesis, the line representing the supposed ancient thalweg appeared in a later coloured version of the map (see Fig. 3.2), based on Delorme Colaço's original black-and-white map, where this line had not been included.

[3] Vasconcelos and Delorne-Colaço's report on the Verdoejo Islands, dated June 1859, has been reproduced by Barreiros (1961–1965, III, no. I-II (1964), 79–81).

In any case, once these disputes had come to an end, Article 1 of the 1864 BT resolved the following: on the one hand, the islands Canosa, Cancela, Ínsula Grande and the islet Fillaboa were awarded to Spain, based primarily on their greater geographical proximity to the right (or Spanish) bank of the Minho river; and on the other, the islands Canguedo and Raña Gallega were awarded to Portugal, based on their greater proximity to the left bank of the river.

3.3.3 The Disputed Criterion for Delimiting the River Border

The question of what general criterion should be followed for tracing the international border of the Minho fluctuated between two possible solutions, which, as mentioned, were already common in treaties relating to river boundary delimitation: either drawing the dividing line along the median line of the river, equidistant between both banks; or following the thalweg, where the current is usually faster and therefore more suitable for navigation. The instructions approved by the Spanish and Portuguese governments in August 1855 for the operation of the JBC (AHN, doc. 3) stipulated that in the border rivers, the commission should preferably use the '*álveo*' (the Portuguese version used the term 'thalweg') to set the international boundary, 'when no tendency to vary is noted in them', something that the commissioners were immediately able to examine upon a detailed inspection of the course of the Minho. In many parts of the river, the line of the thalweg zigzagged noticeably when approaching the edges of either bank, as was evident in the spectacular map (74.5 cm high by 367 cm long) of the stretch between the towns of Caldelas, Tui and Valença do Minho, produced by the engineer Manuel de Uceda at the request of Salvador Tavira, the then-president of the Spanish section of the JBC, in the summer of 1858 (Fig. 3.3).

This map, supplementing the extensive report written by Tavira (AHN, doc. 4), represents the line of the thalweg or the maximum depth of the current river (called 'the new river' on the map), as well as the limit of the old river channel (the 'old river') on the left or Portuguese bank. In some areas, the old riverbed is completely dry and occupied by crops, while in others, it remains active, although largely abandoned for navigation purposes in favour of the new channel. In addition, the map and report provide a wealth of information relevant to the negotiations surrounding the 1864 BT, regarding both the river islands in dispute and the fisheries, jetties and other constructions located along the riverbanks. These are numbered, and the map even includes freehand sketches of some of them. In fact, both documents were commissioned by the Spanish government to study Portuguese fisheries in detail and urge the Spanish border authorities to be on the lookout for any possible new constructions that could hinder navigation or harm Spanish interests. Building on the line of interpretation initiated by Madina and Bourman in 1855, from Spain's point of view, the alteration of the Minho channel due to such works had gradually shifted the river towards the Spanish bank, causing flooding and the loss of crop areas on that side, while also generating land gains on the Portuguese side. The map strives to depict the Minho's changing riverscape, with graphic and textual references not only to the old and new channels and to the land gained (or 'usurped', to use the map's wording) by the Portuguese as a result of these changes but also to the water level in summer and winter.

The 1864 BT, however, was imprecise when referring to the general criterion to be followed in the delimitation of the river boundary. Article 1 of the treaty stipulates that 'the line of separation between the sovereignty of the Kingdom of Spain and that of the Kingdom of Portugal, starting from the mouth of the Minho River [...] shall proceed along the centre of the principal stream of the Minho to the confluence of the river Barjas or Troncoso' (United Nations 1991, p. 276), with the exception of certain islands and islets (such as Canosa or the Verdoejo group) that, as already indicated, were subject to specific solutions. The expression 'centre of the

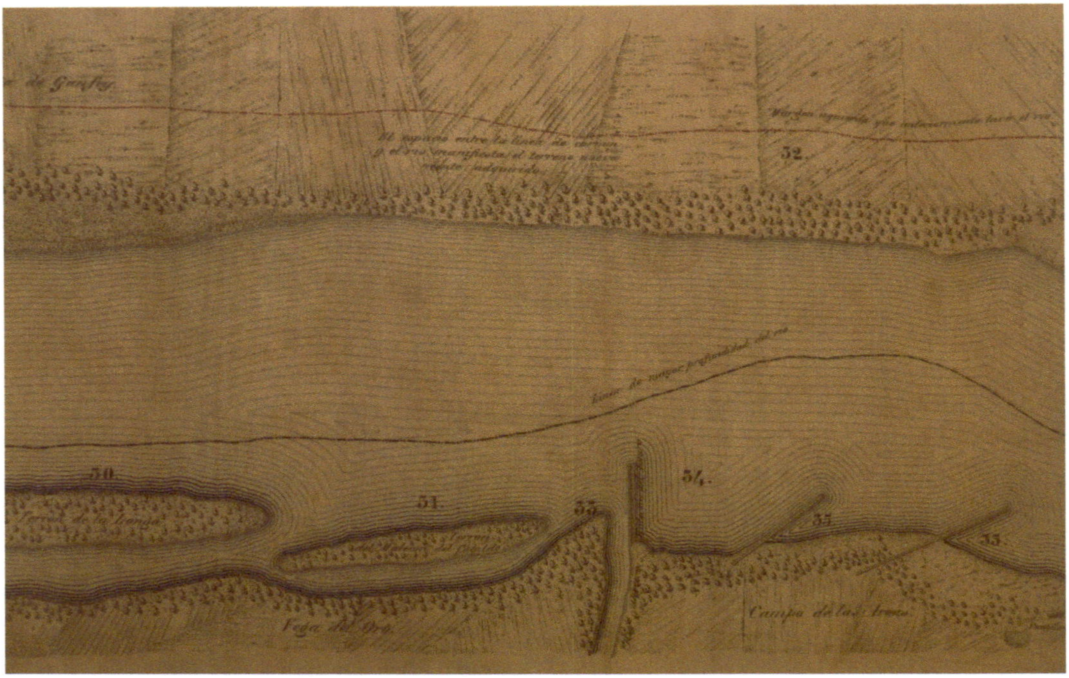

Fig. 3.3 Detail of the *Plano topográfico del río Miño, desde los Baños de Caldelas hasta la barca de pasaje de la ciudad de Tuy a Valenza, en Portugal* [Topographic map of the Minho River, from the Caldelas Baths to the ferry boat crossing from the town of Tui to Valença, in Portugal], prepared [by M. de Uceda] by order of S. Tavira in the summer of 1858. Without scale. Reversing the normal orientation, the right, northern or Spanish bank of the Minho is represented on the lower side of the map, and the left, southern or Portuguese bank on the upper side. From top to bottom, the carmine line drawn on the Portuguese bank represents the southern limit of the old channel on that bank; the dotted black line running along the river is the thalweg or maximum depth of the river; and the numbers at the bottom correspond to natural landforms (islets, sandbanks, etc.) and artificial works (jetties, fisheries, etc.) described in the complementary report of the map. *Source* AHN, Ministry of Foreign Affairs Section, Maps and Charts, 1138

principal stream'—which is not without ambiguity and therefore open to various interpretations—was clarified later as the median line of the river (or central line equidistant from both banks), and not as the thalweg line. This appears in Article 1 of the annex or regulation on boundary rivers approved in Lisbon on 4 November 1866, which states that all frontier rivers included in the 1864 BT belong 'to each of the two nations to the extent of one half of their streams' (United Nations 1991, p.285), as well as in the Act of delivery of the Minho River boundary (Article 2), signed in Lisbon on 30 March 1897, and the General Act of Demarcation, also signed in Lisbon on 1 December 1906, which, in its second part (Article 3), ratifies and finalizes this criterion (*Acta general de demarcación de la frontera entre España y Portugal desde la desembocadura del río Miño hasta la confluencia del Caya con el Guadiana.* Reproduced in *Acuerdos fronterizos*... 2005, pp. 217–333).

3.3.4 The Problem of Artificial River Works

The annex on boundary rivers that accompanied the 1864 BT, despite enshrining the common use of these waters, laid out a series of provisions to regulate this use and address the aforementioned problems caused by the uncontrolled proliferation of fisheries and other constructions along these rivers, especially the Minho. In this regard, Article 4 of the annex prohibited '[the construction], in the rivers or on their banks or those of their islands, [of] works of any kind that might be detrimental to navigation or alter the course of the waters or in any way impair the condition of the rivers with respect to common and public use.' (United Nations, 1991, p. 285). In turn, the transitional article of this same regulation provided for an examination of the frontier portion of the Minho by two engineers (one from each country) in order to verify and describe, among other aspects, 'the obstacles that hinder navigation in various places, the means for removing them and the works which they deem it necessary to construct or demolish, both for making the river navigable and for regularizing the course of its waters' (United Nations, 1991, p. 287). In practice, the effective fulfilment of these technical tasks was finally entrusted to the JBC, which prepared the Act of delivery (*'Acta de entrega'*) of the Minho border line, signed, as already noted, in 1897 under the presidency of Ignacio Salinas-Angulo (colonel of the General Staff of the Spanish Army) and Sebastião Lopes de Calheiros e Meneses (general of the General Staff of the Portuguese Army).

The process of demarcating the border delimited by the 1864 BT lasted for nearly 40 years, although the main part of it was carried out between 1883 and 1906. The demarcation gave rise to two types of legal instruments: partial acts ('*actas parciales*', which describe specific parts of the border) and the aforementioned General Act of Demarcation (which describes the border as a whole), signed in 1906. All these acts were accompanied by a topographical and 'geometric' description and specific cartography of the border line, carried out jointly by the Portuguese and Spanish members of the JBC, which, when attached to these legal documents, also acquired legal status (García-Álvarez and Puente-Lozano 2015b).

In this regard, the JBC produced and published three general cartographic series of the demarcated border by 1906: one at a scale of 1:10,000, another at a scale of 1:50,000, and another at a scale of 1:100,000 (the latter annexed to the General Act of 1906). Furthermore, and exceptionally, the commission produced a *Plano de la línea fronteriza definida por el río Miño* ('Map of the border line defined by the Minho River') at a scale of 1:2500, published in 1898 in a 59-sheet document, which served to complete the aforementioned Act of delivery of the Minho River border line. The latter is another substantial, large-format, manuscript document spanning over 300 pages, which includes a detailed geographical description of this section of the river, as well as an exhaustive inventory of the works and constructions mentioned above, indicating their location, the name of their owner and other observations (AHN, doc. 5). In addition, Article 6 of the act listed the fisheries that 'must be modified or destroyed due to their pernicious influence on the water regime', totalling eleven in all (seven on the right or Spanish bank and four on the left or Portuguese bank).

The map of the Minho—also known as the Minho Atlas (*Atlas del Miño*)—was drawn up by Miguel Correa-Oliver, commander of the Spanish General Staff, and Gaspar Antonio d'Acevedo-Meira, major of the Portuguese General Staff. It is a truly unique and exceptional cartographic document. The map was produced to resolve the disputes affecting relations between both banks and countries with respect to the fisheries, and also to properly manage such works in order to avoid, in the future, the damage associated with their uncontrolled construction. The purpose of this atlas was, therefore, eminently practical, and the choice of such a detailed scale (1:2500) responded to the need to accurately represent the constructions that existed along the course of the river and that could not be properly reflected in the cartography series at scales of 1:10,000 and 1:50,000.

In this way, the preparation of this map or atlas by the two commissioners was done in accordance with other specific instructions prepared by Salinas-Angulo, dated April 1896 (AHN, doc. 6). Furthermore, the details of the process of producing the atlas were collected in an unpublished report, signed in December

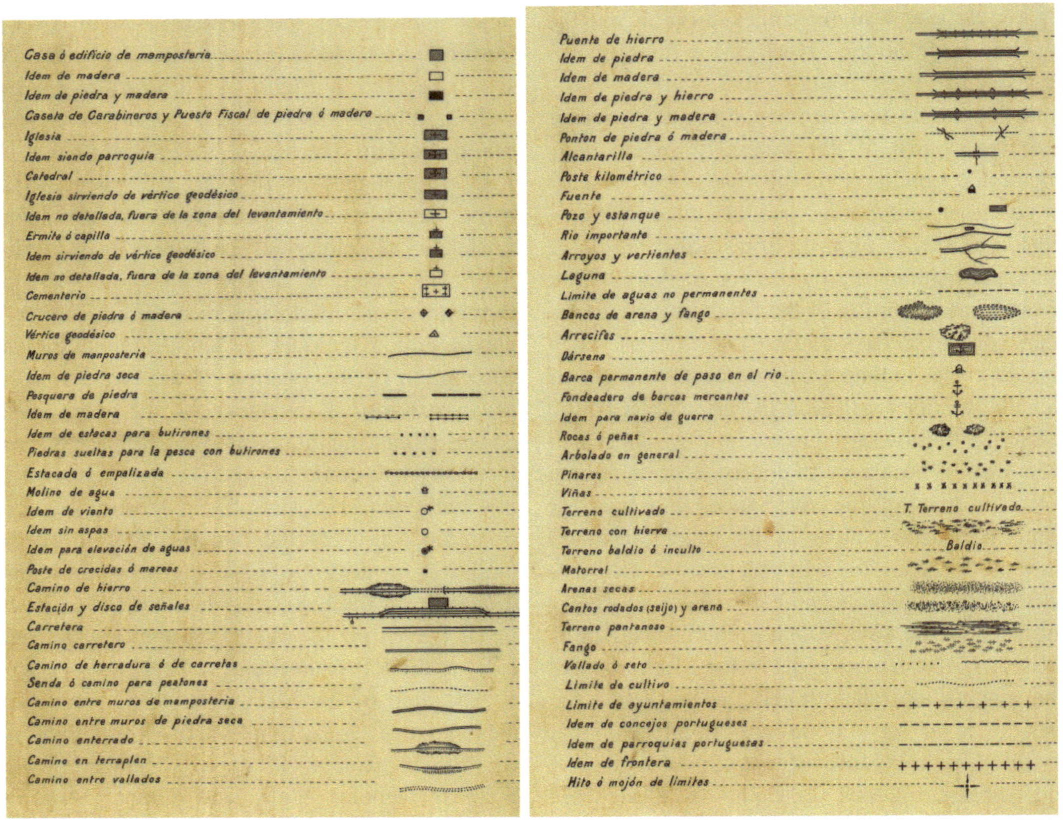

Fig. 3.4 International Boundary Commission between Spain and Portugal: *Plano del río Miño desde su desembocadura hasta la confluencia del río Barjas o Troncoso* [Map of the Minho River from its mouth to the confluence of the Barjas or Troncoso River], produced on a scale of 1:2500 by Miguel Correa-Oliver and Gaspar Antonio d'Acevedo-Meira, 1898. Legend of symbols (Spanish version). *Source* ACEGCGE, Ar. I-T.10-C.2-105bis

1896 by commander Correa-Oliver (AHN, doc. 7). The specific topographical field work to produce the atlas was carried out in the summer of 1896, relying on the Portuguese geodetic network and lasting three and a half months. As already mentioned, the atlas, which has been rightly described as a 'veritable cartographic monument' (Garcia, 2000, p.66), consists of 59 sheets of maps. To these were added the map sheet index, the legend of symbols used (Fig. 3.4), and three folios detailing the list of existing works on both banks, which are 778 in total (including stone fisheries, wooden stockades, docks, quays, retaining walls, boat ramps, etc.). A total of 32 copies were made of the atlas: the two that served for the partial acts and another 30 distributed equally between the Spanish Ministry of State and the Portuguese Ministry of Foreign Affairs, which, in turn, distributed them to various organizations within their respective countries.

The sheets of the atlas represent a wide range of features of geographical interest along the river and its immediate banks, extending up to 250 m on both sides of the bank (Fig. 3.5). In addition to the agreed border line and typical elements found in general topographic maps—such as administrative divisions, the road network and settlements—the maps include information on land uses, the fisheries and other artificial structures, as well as tax and police posts, jetties and anchorages used by passenger, merchant and war ships; the limits of permanent and non-permanent waters; islands

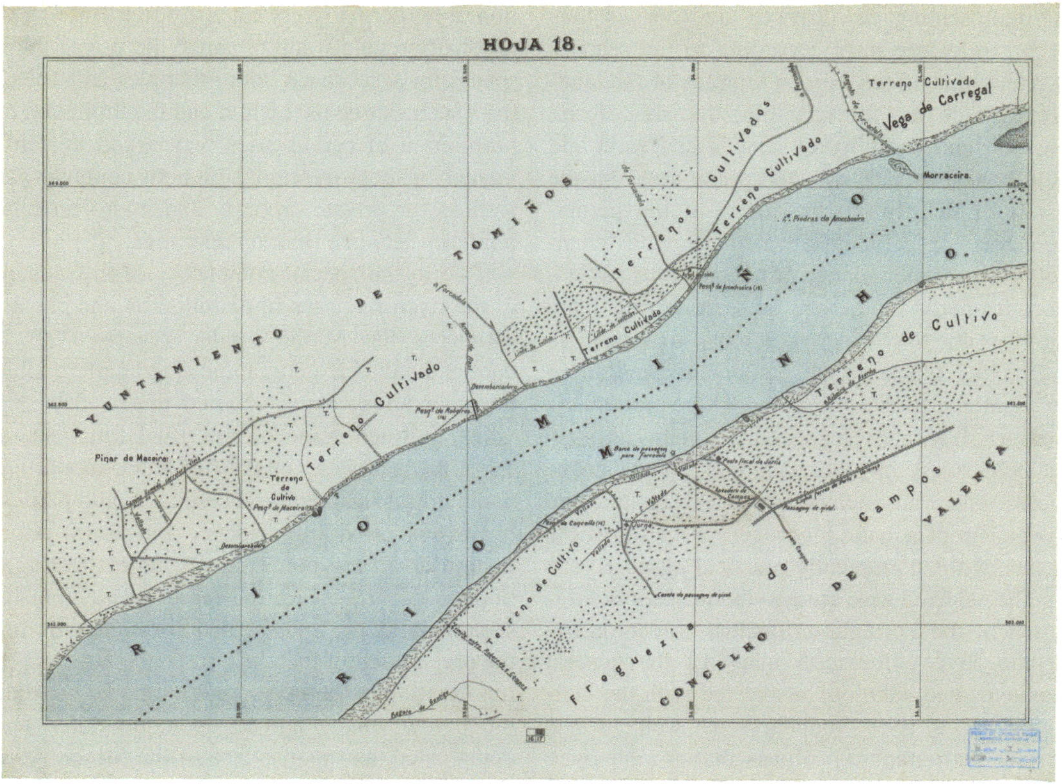

Fig. 3.5 International Boundary Commission between Spain and Portugal: *Plano del río Miño desde su desembocadura...* Sheet no. 18 [Tomiño/Valença]. *Source* ACEGCGE, Ar. I-T.10-C.2-105bis

and sand or mud banks; wetlands; reef areas and the different types of sandy and rocky formations; and so on. All these elements, as previously noted, are detailed in the geographical description that accompanied the deed relating to the delivery of the Minho River.

3.4 Conclusion

Throughout this work, we have reflected on the multiple challenges involved in the delimitation of river borders and how this delimitation is closely intertwined with issues related to the use, management and administration of these boundaries. This contribution has also shed light on how the cartography associated with the delimitation processes of international rivers is not only a more or less faithful reflection of the disputes that arise during these processes but that it also influences these conflicts in a variety of ways: It can spark fresh disputes when maps are inaccurate; fuel them when the maps are used selectively to reinforce one party's position; or help resolve them by providing key geographical information, supporting evidence or simply representations of the finally agreed-upon borders—which, if accompanied by legal texts, acquire legal value in their own right. The challenge of fixing the moving line of river boundaries also entails a cartographic challenge of representing a set of elements, many of them variable, that can be relevant or decisive when it comes to addressing the delimitation, use, and management of these spaces.

Among such elements, it is worth noting the direction of the main currents, the contour lines and depth measurements (fundamental, in turn, for representing the thalweg lines or anchorages), the water level according to the seasons and how this affects certain features of the landscape (such as, for example, the area, form, position and visibility of islands and sand and mud banks, as well as flood-prone areas on the banks of the rivers), the limits of the permanent and non-permanent waters, previous channels currently abandoned or relegated for navigation, formerly submerged lands reclaimed from the river due to channel variations, streams, reefs and rocky obstacles, or settlements and more substantial artificial structures. These particularities turn river maps in general, and maps of international rivers in particular, into unique documents where an attempt is made to capture the dynamism of aquatic landscapes at various time scales, with an eminently practical purpose.

The selected case study—the modern delimitation of the Portuguese-Spanish border on the Minho River—illustrates many of the general problems and solutions associated with the definition of a river boundary, as well as the diverse cartographic practices that are deployed in this type of process. Despite being a 'natural' border of great antiquity, the demarcation of this stretch became one of the most controversial issues in the process of delimiting the Spanish–Portuguese border, and it ended up being decisive for the resolution of a number of key issues related to the management and demarcation of the rest of the bordering rivers affected by the 1864 BT. The commissions responsible for delimiting the Minho border had to deal with, among other aspects, disputes related to the sovereignty of some river islands, the problem of defining a dividing line in a channel affected by high variability, or the proliferation of uncontrolled artificial constructions that altered the course of the river and caused or contributed to various harmful repercussions with respect to navigability, the use of land and certain agro-pastoral and fishing resources, the persecution of smuggling, and military defence and security. The 1864 BT and the complementary regulations that supplemented it provide an arsenal of essential instruments not only for the delimitation of bordering rivers but also for attempting to rationally regulate and organize the use of these spaces, in accordance with principles and objectives such as free navigation and the limitation or prohibition of certain uses understood as detrimental to the sovereignty of both countries, as well as for private property and, more broadly, relations between riparian countries.

Due to the special complexity of this stretch of the river, the tasks of delimitation and demarcation of the Minho border generated abundant and valuable geographical and cartographic documentation, barely studied until today, and which fulfilled some of the usual functions in this type of process: on the one hand, supporting the interests and territorial claims of both parties, by way of providing evidence and useful information about the border areas subject to litigation, as occurred with the representations made separately by the Spanish and Portuguese during the preparation of the 1864 BT; and on the other, representing in the most precise manner possible the border finally agreed upon, delimited and demarcated, as was the case of the Minho Atlas at a scale of 1:2500.

These representations are, in short, of enormous interest, both regarding historical geography and the history of geography and cartography. With regard to the latter, their value is multiple. First, they allow us to examine military and diplomatic thinking about a given territory, as well as its representation techniques. Second, they offer a broad repertoire of the languages and codes of river cartography in general, and of international river cartography in particular—these representations are also inseparable from the formation of the modern concept of border. Third, the maps of the case analysed here constitute very useful sources for understanding the landscape and spatial organization of the border areas represented during the period in which their delimitation and demarcation took place. Finally, it should be noted that much of this documentation, especially the cartography produced during the demarcation phase subsequent to the 1864 BT, was the result of collaboration between the military commissioners of both countries, which is why its analysis allows progress in the development of an international and transnational perspective on the history of cartography and geographical knowledge.

Archival Sources

ACEGCGE. Archivo Cartográfico y de Estudios Geográficos del Centro Geográfico del Ejército. Madrid.

AHDMNE. Arquivo Histórico-Diplomático do Ministério dos Negócios Estrangeiros. Lisbon.

AHN (Archivo Histórico Nacional/National Historical Archive, Madrid), doc. 1: Report sent from Bourman to the Spanish government on 5 October 1855 (AHN, Ministry of Foreign Affairs Section, Treaties, Tr. 138, Negotiations 207, File 20).

AHN, doc. 2: Instructions given to Bourman, dated 17 October 1855, in reply to the report sent by him on 5 October 1855 (AHN, Ministry of Foreign Affairs Section, Treaties, Tr. 138, Negotiations 207, File 20).

AHN, doc. 3. "Instrucciones para la Comisión Mixta encargada de la demarcación de la frontera entre los Reynos de España y Portugal" ("Instructions to the Mixed Commission in charge of the demarcation of the border between the Kingdoms of Spain and Portugal"), (AHN, Ministry of Foreign Affairs Section, Treaties, Tr. 138, Negotiations 207, File 20). The Portuguese version can be seen in Barreiros (1961–1965), I, no. I-II (1961), 110–111.

AHN, doc. 4. "Observaciones descriptivas e históricas para la completa inteligencia del plano topográfico del río Miño, desde Caldelas al sitio de la barca del pasage, de la ciudad de Tuy a Valença, según se hallaba en el verano de 1858" ("Descriptive and historical observations for the complete understanding of the topographical plan of the river Minho, from Caldelas to the site of the ferry boat from the city of Tuy to Valença, as it was in the summer of 1858") (AHN, Ministry of Foreign Affairs Section, Treaties, Tr. 138, Negotiations 207, File 15).

AHN, doc. 5: *Acta relativa a la entrega del río Miño que sirve de frontera entre España y Portugal desde su desembocadura hasta la confluencia del río Barjas o Troncoso.* AHN, Ministry of Foreign Affairs Section, Treaties, Tr. 152, Negotiations 224, File 8), This document has been partially reproduced in García-Mañá (1993, pp. 251–289).

AHN. doc. 6: "Instrucciones que han de observarse para levantar las orillas del río Miño, en escala 1:2500" ("Instructions to be observed for surveying the banks of the river Minho, on a scale of 1:2500') (AHN, Ministry of Foreign Affairs Section, Treaties, Tr. 154, Negotiations 224, File 20).

AHN, doc. 7: "Memoria sobre los procedimientos empleados para el levantamiento de la porción fronteriza del río Miño en la escala de 1:2500" ("Report on the procedures used for the survey of the border portion of the river Minho at a scale of 1:2500"), signed at Tui by M. Correa-Oliver, 22 December 1897 (AHN, Ministry of Foreign Affairs Section, Treaties, Tr.154, Negotiations 224, File 20).

References

Acuerdos fronterizos con Portugal y Francia. Colección de recopilaciones (2005). Ministerio de Asuntos Exteriores y Cooperación, Madrid, vol 1

Barreiros JB (1961–1965) Delimitaçao da fronteira lusoespanhola. O Distrito de Braga, I (I–II), 1961: 59–148: I (III–IV), 1961: 335–413; II (I–II), 1963: 83–170; II (III–IV), 1964: 81–176; III (I–II), 1964: 1–96; III (III–IV), 1965: 401–489

Biger G (1988) Physical geography and law: the case of international river boundaries. GeoJournal 17(3):341–347

Biggs M (1999) Putting the state on the map: cartography, territory, and European state formation. Comp Stud Soc Hist 4(2):374–411

Bishop P et al (2018) River boundaries guidebook, African Union

Black J (2009) Change in Ancien Régime international relations: diplomacy and cartography 1650–1800. Diplomacy Statecraft 20:20–29

Blake G (1995) The depiction of international boundaries in topographic maps. IBRU Boundary Secur Bull 3(1):44–50

Bouchez LJ (1963) The fixing of boundaries in international boundary rivers. Int Comp Law Q 12:789–817
Branch J (2013) The cartographic state: maps, territory, and the origins of sovereignty. Cambridge University Press, Cambridge
Carroll FM (2004) The search for the Canadian-American boundary along the Michigan frontier, 1819–1827. Mich Hist Rev 30(2):77–104
Caflisch L (1989) Les frontières interétatiques se rattachant aux cours d'eau internationaux. Recueil des Cours De L'academie De Droit International De La Haye 219(7):62–103
Costa MF (2017) Ver y no publicar, publicar sin conocer: historias sobre la representación del río Madeira. Hist Mex 67(2):819–858
Dodds K (2021) Border wars: the conflicts that will define our future. Ebury Press, London
Donaldson JW (2009) Where rivers and boundaries meet: building the international river boundaries database. Water Policy 11(5):629–644
Donaldson JW (2010) Marking Territory: demarcation of the DRC-Zambia boundary from 1894 to the present day. Doctoral Thesis, Durham University, Durham
Donaldson JW (2011) Paradox of the moving boundary: legal heredity of river accretion and avulsion. Water Alter 4(2):155–170
Duinat B (2022) A water borderline adrift between France and Spain: the Bidasoa and the Bay of Higuer at the end of the nineteenth century. In: García-Álvarez J, Puente-Lozano P (eds) Beneath the lines: borders and boundary-making from the 18th to the 20th century. Springer, Cham, pp 125–138
Garcia JC (2000) A fronteira impressa: apontamentos sobre uma série cartográfica. In: López Trigal L, Guichard F (eds) La frontera hispano-portuguesa: nuevo espacio de atracción y cooperación. Fundación Rei Afonso Henriques, Zamora, pp 57–72
García-Álvarez J, Puente-Lozano P (2015a) La delimitación de la frontera luso-española en el río Miño (1855–1906). Conflictos territoriales y representaciones cartográficas". Terra Brasilis (Nova Série) 6 https://doi.org/10.4000/terrabrasilis.1708
García-Álvarez J, Puente Lozano P (2015b) Las Comisiones Mixtas de Límites y las representaciones geográficas de la frontera hispano-portuguesa (1855–1906). Revista de Historiografía 23:67–100. https://erevistas.uc3m.es/index.php/REVHISTO/article/view/2756
García-Álvarez J, Puente-Lozano P (2016) La frontera hispano-portuguesa en el río Miño en el Tratado de Límites de 1864. Deslinde y ordenación de un territorio fluvial conflictivo. In: Vera F et al (eds) Paisaje, cultura territorial y vivencia de la geografía. Universidad de Alicante, Alicante, pp 761–787
García-Álvarez J, Puente-Lozano P (2017) A controversial natural border. The making of the Spanish–Portuguese boundary along the Minho River (1855–1906). Geoforum 79:5–16
García-Mañá LM (1993) Miño: ¿existiu una fronteira? Galaxia, Vigo
Giménez P (2009) Cartografía histórica y alteración de cursos fluviales en la España mediterránea. In: Alberola A, Olcina J (eds) Desastre natural, vida cotidiana y religiosidad popular en la España moderna y mediterránea. Universidad de Alicante, Alicante, pp 391–432
Gleditsch NP et al (2006) Conflicts over shared rivers: resource scarcity or fuzzy boundaries? Polit Geogr 25(4):361–382
Grainger S, Conway D (2014) Climate change and international river boundaries: fixed points in shifting sands. Wires Clim Change 5:835–848
Jones SB (1945) Boundary-making: A handbook for statesmen, treaty editors and boundary commissioners. Carnegie Endowment for International Peace, Washington, DC
Prescott V, Triggs G (2008) International frontiers and boundaries. Law, politics and geography. Martinus Nijhoff, Leiden
Sáry P (2019) Water law rules in Ancient Rome. J Agric Environ Law XIV(26):219–254
Schröter F (1992) Le système de délimitation dans les fleuves internationaux. Annuaire Français De Droit International 38:948–982
Shelburne OB (2020) From Presidio to the Pecos River: surveying the United States-Mexico boundary along the Rio Grande, 1852 and 1853. University of Oklahoma Press, Oklahoma
United Nations (1991) Treaty series. Treaties and international agreements registered or filed and recorded with the Secretariat of the United Nations, vol. 1288. United Nations, New York. Retrieved from: https://treaties.un.org/doc/Publication/UNTS/Volume%201288/v1288.pdf
Van den Brink P (2000) River landscapes: the origin and development of the printed river map in the Netherlands, 1725–1795. Imago Mundi 52:66–78
Warmoes I et al (eds) (2016) Cartographie et Traités de Paix (XV-XXe siècle). Cartes et Géomatique 228:10–182
Zusman P, Nunes S (2022) Borders on the paper, borders on the ground: boundary commissions under the Treaty of Madrid and their role in the reinvention of the Rio de la Plata frontier (1752–1759). In: García-Álvarez J, Puente-Lozano P (eds) Borders and boundary-making from the 18th to the 20th century. Springer, Cham, pp 93–108

4 Maps, the Linearization of the Hydrographic Border, and the Reconfiguration of Power: The Demarcation of the International Boundary Line Between Argentina and Brazil (1894–1910)

Perla Zusman and Sergio Nunes-Pereira

Abstract

At the turn of the twentieth century, Argentina and Brazil, were engaged in struggles over Latin America hegemony. During this period, the United States not only organized the first Pan-American Conference (1889–1890) but began positioning itself as an arbitrator in border disputes. In 1894 Argentina and Brazil appointed the US president, Grover Cleveland, to mediate in their territorial dispute over the region known as Misiones (by Argentina) or Palmas (by Brazil). After a treaty between Argentina and Brazil was signed in 1898, confirming the US arbitral award, the two countries took steps to implement on the ground the decisions that had been taken on paper. This chapter identifies strategies used by the demarcators—practices, techniques and instruments they employed—to address the challenges of delimiting international borders along watercourses. In other words, their task was to impose a geometric order on the frontier along the Uruguay, Pepiri-Guazu, San Antonio and Iguazu rivers. In this context, cartography not only expressed the negotiations results on the ground, it also depicted a continuous boundary line, helping anchor the territorial imaginary of demarcation. The consensus between Argentina and Brazil on the hydrographic boundary contributed to ensuring friendly relations in the early twentieth century.

Keywords

Argentina-Brazil relations · Hydrographic border · Cartopolitics · Joint boundary commissions · Misiones/Palmas dispute · Diplomatic cooperation · Thalweg · Parana river · Iguazu river · Iguazú falls

4.1 Border Conflicts and Establishing Regional Dominance

Between the middle of the nineteenth century and the early decades of the twentieth, the countries of South America transformed themselves into modern states. From a territorial point of

This work has been supported by the Spanish State Research Agency under Grant Number PID2020-114088GB-I00/AEI/10.13039/501100011033.

P. Zusman (✉)
CONICET and University of Buenos Aires, Buenos Aires, Argentina
e-mail: perlazusman@yahoo.es

S. Nunes-Pereira
Fluminense Federal University, Niterói, Brazil
e-mail: snunesrio2@gmail.com

J. García-Álvarez and A. Reyes Novaes (eds.), *Mapping the Contested Line*, Historical Geography and Geosciences, https://doi.org/10.1007/978-3-032-29889-8_4

view, this meant appropriating territories that had been in the hands of Indigenous peoples and, at the same time, delimiting their borders with neighbouring countries. These tasks were accompanied by strategies intended to provide the territory over which dominance was being asserted with unity and cohesion at the material level, by extending communications infrastructure such as railways and the telegraph and by producing maps, and at the symbolic level, by constructing the imagined community of the nation (Magnoli 1997; Oszlak 2004; Lois 2019). These processes involved struggles between the developing states to gain dominance within the region. Based on Carl Von Clausewitz's and Alfred Thayer Mahan's principles, it was believed that there was a relationship between the control of land, rivers and seas and gaining dominance. The methods by which international boundary conflicts were resolved involved the possibility that the land, river or sea areas controlled by each state might be increased or decreased. While in some cases border conflicts led to military confrontations (such as the War of the Pacific between Chile and Peru, 1879–1883, and the Acre War between Bolivia and Brazil, 1899–1903), in others they gave rise to races to acquire arms or naval power (Garay 2012; Jiménez Cabrera and Garay Vera 2021). Garay (2012) understands these races as a competition between contenders seeking to acquire increasingly technologically advanced war matériel as a way of protecting themselves from the threats posed by neighbouring countries. While Argentina and Chile engaged in an arms race during the 1890s and early 1900s, driven by border disputes involving maritime zones of influence,[1] the arms race between Argentina and Brazil was precipitated by Brazil's acquisition of cutting-edge naval equipment between 1906 and 1914 (Martins Filho 2023; Ventura de Marco 2023). This militaristic approach coexisted alongside others seeking rapprochement and conciliation, not only between Argentina and Brazil but between all the states in the Southern Cone (Doratioto 2014; Matus Matus and Aranda Bustamante 2016).

At the same time, the United States joined in the competition for empire, with the intention of turning Latin America into a sphere of commercial and territorial expansion. The inter-American conferences held from 1889 onwards gave shape to the US Pan-Americanist project, which sought to become the guardian of the region and prevent any European intervention (Mc Gann 1957; Cervo and Bueno 1992; Morgenfeld 2011). US military intervention in countries such as Cuba (in 1898) and Panama (in 1903) led to the Pan-Americanist project being interpreted as an attempt to destroy the independence of the region's nation states. While Argentina's liberal elites were unwilling to set aside their relations with European countries in favour of trade with the United States, Brazil's elites promoted rapprochement with the US (Bandeira 2004, 2010).[2]

One of the topics on the agenda at the first Pan-American Conference of 1889–1890 was the issue of arbitration. The principle of arbitration was proposed as a way of resolving territorial disputes and those over diplomatic and consular privileges, indemnities, navigation rights, intelligence and the enforcement of treaties (Chianelli 1989).[3] The United States thus sought to set itself up as the mediator of disputes par excellence and the guarantor of peace. Its strategy of prioritizing diplomatic solutions over those involving military confrontation added to the set of 'virtues' that some Latin American leaders perceived in the US political model—a democratic, republican,

[1] The signing of the 'May Pacts' in 1902 not only contributed to the delimitation of the international border between Chile and Argentina, but also set limits on arms purchases by each signatory for five years (Lacoste 2001; Ventura de Marco 2023).

[2] In the context of the Proclamation of the Republic in Brazil, and while the first Pan-American Conference was being held in Washington, the new Brazilian government started to forge closer links with the US. In 1891, Brazil and the US signed a reciprocal trade agreement that provided for differentiated treatment of specific Brazilian and US products, mainly affecting trade in coffee and wheat (Sotomayor 1996).

[3] Our translation. Unless otherwise stated, all translations of cited foreign language material in this chapter are our own.

egalitarian country where individual effort guaranteed social advancement. At the same time, US participation in arbitration processes was intended to displace Great Britain from this role in the region and from its hegemonic position over the region within the world system (Wallerstein 2004).

Examples of these interventions include Grover Cleveland, president of the United States from 1885 to 1889 and from 1893 to 1897, acting as arbitrator in the conflict between Costa Rica and Nicaragua over the control of the territory of Guanacaste (in 1888); his mediation in the conflict between Venezuela and Great Britain over the territory of Guayana Esequiba (in 1897); and William Insco Buchanan acting as Grover Cleveland's representative in the dispute between Chile and Argentina over the Puna de Atacama (in 1899) (Zusman and Hevilla 2014).

In the view of some historians, the attempts to bring about closer diplomatic relations between Argentina and Brazil and move away from military confrontation were clearly linked to the US's Pan-American project to exert economic influence in the region. This diplomatic approach would achieve greater prominence towards the end of the first decade of the twentieth century and would lead to the signing in 1915 of the Consultation, Non-Aggression and Arbitration Pact (ABC Pact) by Argentina, Brazil and Chile (Doratioto 2014; Matus Matus and Aranda Bustamante 2016).

This was the regional political context in which Argentina and Brazil had to attempt to resolve the Misiones/Palmas dispute. The territory in question, covering 19,000 square kilometres, was located in the far east of Misiones Province, according to Argentina's claim, but according to Brazil, was part of the Brazilian state of Paraná (southern Brazil), under the designation 'Comarca de Palmas' (jurisidiction of Palmas) (see Fig. 4.1). The disputed area was important for Brazil, as in the Brazilian geopolitical imaginary, it served as a buffer against the isolation of Rio Grande do Sul.

According to the contemporary documentation, the dispute was the result of a lack of agreement between the joint commissions parties appointed by the Spanish and Portuguese empires in implementing the provisions of the Treaty of San Ildefonso (1777). The principles of *uti possidetis juris* and natural borders were not enough to ensure that the border could be drawn clearly, accurately and permanently. The joint boundary commission failed to identify clearly on the ground the Pepiri-Guazu and San Antonio rivers upon which the international border was meant to be based. Nor could the difficulties in determining the boundary be overcome during the pre-negotiation processes that took place between 1857 and 1859.

Towards the end of the nineteenth century, the Argentine government's administration of the territory of Misiones (previously under the control of the province of Corrientes), together with the occupation of the disputed area by Brazil and the establishment of military colonies there, prompted negotiations in 1881 to resolve the conflict (Zusman et al. 2023). To this end, in 1885 a treaty was signed that provided for the organization of a joint commission for the exploration and study of the disputed area. However, this demarcation expeditionary party, which worked in the field between 1886 and 1888, was unable to agree on which watercourse and its counter-slope should be used to determine the international border.[4]

While Brazil was in favour of setting the boundary further west, Argentina believed that the Pepiri-Guazu and San Antonio rivers were actually to be found in the contemporary courses of the Chopim and Chapecó rivers. As a result, new diplomatic negotiations were started, which led to the Treaty of Montevideo (1890). This proposed that the territory should be divided into two equal parts. Brazil's Congress rejected this treaty which led to US President Grover Cleveland being requested to arbitrate in 1894. The arbitral award was favourable to Brazil, establishing the borders as running along the Pepiri-Guazu and San Antonio rivers. In 1895 and 1898, both countries signed agreements ratifying

[4] Actas de Conferencias-1886-87-88. [Conferences Reports-1886-87-88] (Argentina, Ministerio de Relaciones Exteriores y Culto 1910, Vol. II: 147–165).

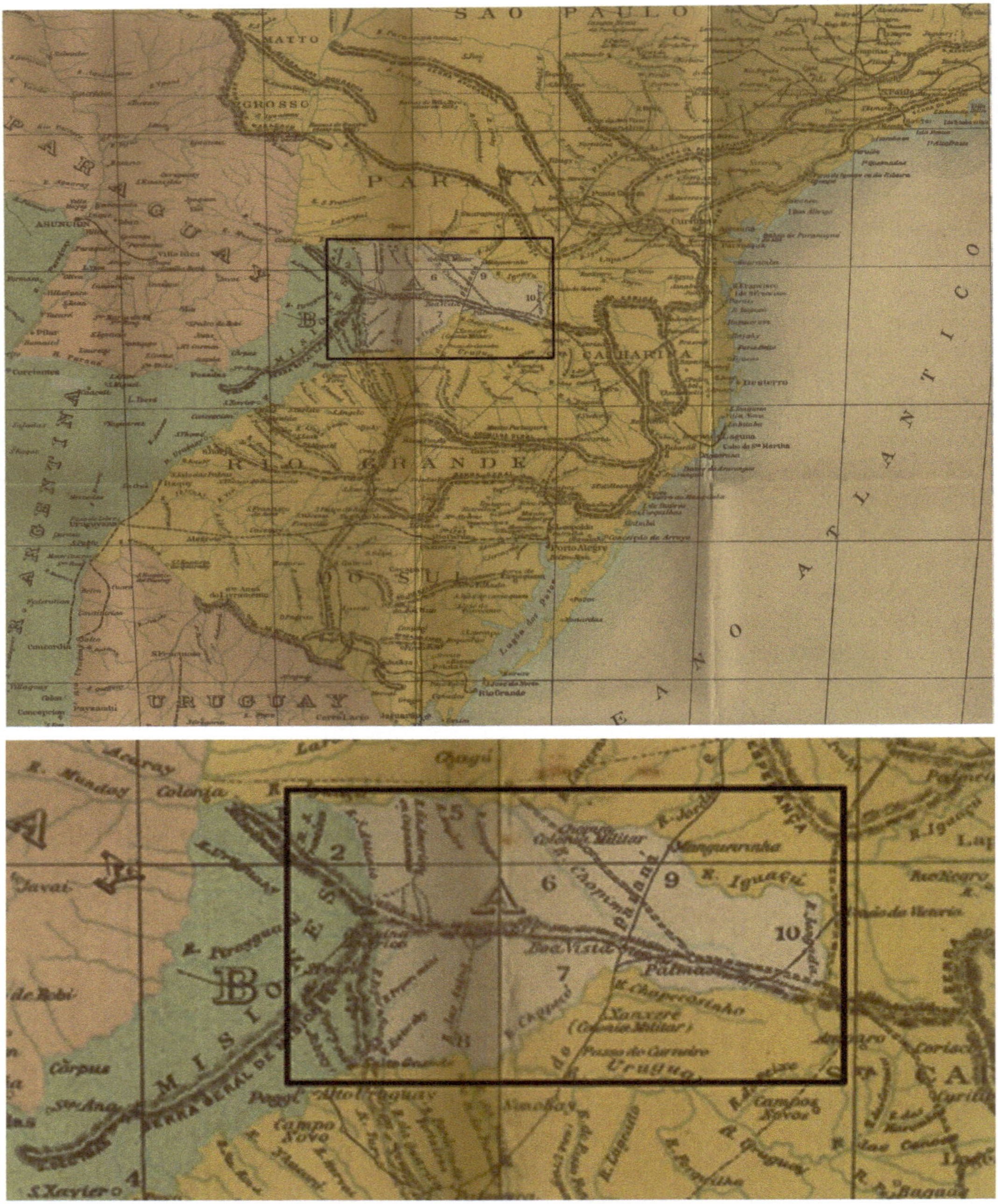

Fig. 4.1 Disputed territory of Misiones/Palmas, as presented by the Brazilian government to defend its claims in the arbitration proceedings before President Grover Cleveland (1894). *Source* Brazil. Ministério das Relações Exteriores (1945) 'Mappa do Brazil Meridional mostrando a parte do seo territorio reclamada pela República Argentina'. References: A: Territory claimed by the Republic of Argentina. This includes part of the jurisdiction of Palmas, State of Paraná, United States of Brazil. B: Territory of Misiones, occupied by Argentina after the War of the Triple Alliance. 1–4: Boundary between Brazil and Argentina, according to Brazil's claim. 6–7: Eastern boundary of the area claimed by Argentina before 1888. 9–10: Eastern boundary of the area claimed by Argentina from 1888 onwards

the arbitral award and establishing the characteristics of the work to be undertaken on the ground, which began in 1900 and continued until 1905. The work was ratified in the 1910 Treaty of Rio de Janeiro.

In this chapter, our goal is first to identify the distribution strategies used by the joint Argentine-Brazilian commission by analysing the practices, techniques and instruments it used to implement the agreements derived from Grover Cleveland's 1895 arbitral award regarding the Misiones/Palmas dispute. Second, we seek to link these demarcation efforts with the intention of preventing any dispute from escalating into armed conflict.[5] In this regard, these activities of distribution between the territories of the two nations can be understood as having contributed to ensuring harmonious and friendly diplomatic relations between Argentina and Brazil, at least for a limited period.

We obtained the sources on which this chapter is based from the archives and libraries of the countries involved in the demarcation. In Argentina, these included the Archivo del Ministerio de Relaciones Exteriores (Ministry of Foreign Affairs Archive); the Archivo General de la Nación (General Archive of the Nation); the Biblioteca Nacional de Argentina (National Library of Argentina); and the Academia Nacional de la Historia de la República Argentina (National Academy of the History of Argentine Republic). In Brazil, our sources included the Archivo del Ministerio de Relaciones Exteriores de Brasil-Itamaraty (AMREBI) (Archive of the Brazilian Ministry of Foreign Affairs, at the Itamaraty Palace) and the Biblioteca Nacional do Brasil (BNB) (Brazilian National Library). Our objective in using resources from both countries was to transcend methodological nationalism (Wimmer and Glick Schiller 2002) and construct a perspective that brings into dialogue—and sets into contrast—not only the documentation produced by the joint Argentine-Brazilian border demarcation commission, but also by each country's respective demarcation commissions.

Many of these sources are cartographic. Taking a cartographic approach allowed us to identify the practices, techniques and instruments used by the demarcation commissions to tame the natural environment and geometrize the border along watercourses, particularly in the Uruguay and Iguazu rivers. By using the maps, we were able to identify the processes by which the commissions determined the thalweg and established which islands belonged to each nation. The mapping process became an expression of cartopolitics, 'where questions pertaining to sovereign control over space are decided through cartography and law' (Strandsberg 2017, p. 818). The cartography thus not only expressed the results of the negotiations on the ground (between the members of the joint Argentine and Brazilian commission, the technical instruments and the natural environment), it also depicted a continuous boundary line, helping anchor the territorial imaginary of demarcation. In some instances, the mapmakers attempted to lend greater texture to the process of cartographic geometrization of the border by providing photographs of the landscapes around the waterfalls or by mapping the border markers.

This chapter is part of a tradition of geohistorical research studies examining the role of demarcation commissions in establishing boundaries. These studies have addressed the complex nature of this process, identifying multi-scale relationships and covering the multiple actors involved in the process of boundary-setting (Di Fiore 2017; Donaldson 2002; García Álvarez and Puente-Lozano 2022). Some of them have had a particular focus on imaginative geographies (García Álvarez and Puente-Lozano 2025) or on the resistance and autonomy of local populations (Sahlins 1990; Herzog 2015; Erbig 2020) while others have focused more closely on the role of the environment, techniques and cartography in redefining and negotiating diplo-

[5] Some of the military events that took place in Latin America as a result of border disputes are still present in the collective memory. This is true of the War of the Pacific (1879–1884), between Chile and the allied forces of Peru and Bolivia, and the Acre War (1899–1903), between Bolivia and Brazil. In both Argentina and Brazil, the press and nationalist circles floated war scenarios between the two countries, a prospect backed by the military and part of the political class.

matic agreements (Zusman 2017; Tamayo Pérez 2022; Urteaga 2022; Zusman and Nunes-Pereira 2022; Altic 2023). All of these studies have attempted to politicize, de-essentialize and show the conflictive nature of these processes.

This chapter will be divided into an introduction and five sections. In the first of these sections, we present the instructions that were given to direct the demarcation work and the composition of the joint commissions. In the second, we describe the challenges of demarcation efforts along the Uruguay River and the implications of choosing the thalweg principle to determine the border line and allocate the islands to one country or the other. The third section focuses on the challenges involved in determining the border line in a rugged environment like the Iguazu Falls. Here, the technical instrument that made it possible to resolve certain cartographic problems faced by the commission and produced photographs of the border landscapes was the phototheodolite. The fourth section describes the practices, techniques and instruments involved in taking coordinates, which made it possible to place the border markers, particularly those that were erected at watersheds, river confluences and on islands. Finally, we present our conclusions, emphasizing the link between demarcation activities and political and diplomatic interests.

4.2 Demarcation Work Along the Uruguay River

After President Grover Cleveland's arbitration ruling, the governments of Argentina and Brazil signed a series of agreements and protocols in order to be able to proceed with the demarcation of the border.

The instructions for carrying out the demarcation proposed the creation of a joint Argentine and Brazilian commission.[6] While General Dionísio Cerqueira headed up the Brazilian commission, the Argentine commission was to be led by Pedro Ezcurra, who was minister of agriculture at the time. Both these officials played an active role in various stages of the construction of their respective state territories, particularly in the demarcation of borders. After taking part in the War of the Triple Alliance, Cerqueira, who was a military and civil engineer from the Military School in Rio de Janeiro, went on to become a member of the joint Brazil-Venezuela boundary commission (1879–1885) and later joined the joint Argentine and Brazilian demarcation commission of 1886. He was also part of the diplomatic mission sent to the United States to prepare the documentation that would be made available for the 1898 Cleveland arbitration (Nascimento 2015). Ezcurra, who was an engineering graduate from the University of Buenos Aires, took part in the exploratory work associated with determining Argentina's boundary with Chile (1889–1902). He was also involved in designing a local railway line in the central region of Argentina (joining Dique San Roque and Cruz del Eje) and mapping the Patagonian governorate of Chubut (1893) created after the appropriation of lands under Indigenous rule (Cutolo 1969).

The other members of the Argentine expeditionary party included a second commissioner (Lieutenant José Moneta), two assistants (Pedro Padilla and Frigate Lieutenant Guillermo Mulvany), a secretary (first Ramón R. Sardá and then Carlos Alberto de Zuviría), two technical assistants (Frigate Ensigns Carlos Ribero and Carlos Valladares) and a navy surgeon (Norberto Pérez). The Brazilian party also included a second commissioner (Major Gabriel Pereira de Souza Botafogo, of the Corps of Engineers), as well as two assistants (Captain Benjamim Liberato Barroso and Captain Alipio Gama), a secretary (Artillery Captain José Leandro Caval-

[6] Tratado con el Brasil para completar por medio de acuerdo amigable y directo el establecimiento de la línea divisoria de los dos países, en parte determinada por el laudo arbitral del Presidente de los Estados Unidos de América, art. 5, art. 6. [Treaty with Brazil to complete, through a friendly and direct agreement, the establishment of the boundary line between the two countries, partly determined by the arbitral award of the President of the United States of America, article 5 and 6] (Argentina, Ministerio de Relaciones Exteriores y Culto 1910, Vol. II, p. 137).

canti), a technical assistant (Otto Max Guichard), a doctor (Joaquim Antonio da Cruz) and a pharmacist (José B. Cepuciro da Costa Reis) (Argentina, Ministerio de Relaciones Exteriores y Culto 1910, vol. I, pp. 526–527). On both sides, most of the more senior members of the commission were army or naval officers, who undertook technical tasks. In addition to these individuals, there was also a contingent of soldiers and workers who performed manual tasks and whose names are not known. Cerqueira stated that the soldiers acted as escorts for the demarcators, set up the camps and steered the boats[7]; he also acknowledged the work of *peones* (labourers) who took part in building canoes, opening paths and clearing land.[8]

The new joint Argentine and Brazilian commission began its work in January 1901, determining geographical positions along the Uruguay, Pepiri-Guazu, San Antonio and Iguazu rivers, in addition to the stretches between these rivers that made up the much shorter land border (it comprised only 2 per cent of the total length of the border). The rivers in question were in the Uruguay and Paraná basins and had a discharge regime characterized by two seasonal high flows that alternated with slightly lower flows (Santos 1977), thus maintaining a significant volume of water throughout the year; they also had specific characteristics that varied according to the slope of the relief, the natural features of their beds and the discharges of their tributaries.

According to the treaty signed by Argentina and Brazil, the demarcation of the dividing line would begin at the Uruguay River, 'at the confluence with the Cuareim River, and continue along the thalweg of the said river up to the mouth of the Pepiri-Guazu River.'[9] From the mouth of the Pepiri-Guazu, the boundary would follow along this river's thalweg to its main headwaters; from there, the boundary would continue along the highest ground to the main headwaters of the San Antonio River and then along the San Antonio's thalweg until it reached the confluence of the San Antonio and Iguazu rivers. From that point it would follow the thalweg of the Iguazu River until the latter reached its confluence with the Paraná.[10]

At that time, international law recognized four valid approaches to fixing boundaries on river courses:

(1) fixing the demarcation lines on both banks of the river. This assumed that the watercourse was *res communis*, i.e., it belonged equally to both states;
(2) determining the delimitation based on just one of the banks of the river. The river was thus under the control of only one of the riparian states;
(3) fixing the demarcation along a midline, equidistant between the closest reference points or between the two banks of the river; and
(4) the delimitation following the thalweg line, which could be understood as meaning the main navigation channel followed by river vessels when descending the river, or as the line connecting the deepest points of the riverbed (Bouchez 1963; Schröter 1992; Caflisch 1989).

---

[7] AMREBI. Comissão de Limites com a República Argentina. File n. 24, May 22 1902 (Box 434, bandle 3); file n. 47, May 14 1903 (Box 434, bundle 3).

[8] AMREBI. Comissão de Limites com a República Argentina. File n. 53, October 14 1903 (Box, 434, bundle 3).

[9] Tratado con el Brasil para completar por medio de acuerdo amigable y directo el establecimiento de la línea divisoria de los dos países, en parte determinada por el laudo arbitral del Presidente de los Estados Unidos de América, artículo 1.[Treaty with Brazil to complete, through a friendly and direct agreement, the establishment of the boundary line between the two countries, partly determined by the arbitral award of the President of the United States of America, article 1] (Argentina. Ministerio de Relaciones Exteriores y Culto 1910, vol II. p. 136).

[10] Tratado con el Brasil para completar por medio de acuerdo amigable y directo el establecimiento de la línea divisoria de los dos países, en parte determinada por el laudo arbitral del Presidente de los Estados Unidos de América, artículo 2 y 3. [Treaty with Brazil to complete, through a friendly and direct agreement, the establishment of the boundary line between the two countries, partly determined by the arbitral award of the President of the United States of America, article 2 and 3]. (Argentina, Ministerio de Relaciones Exteriores y Culto 1910, vol. II, pp 136–137).

As international trade expanded in the nineteenth and twentieth centuries, the thalweg principle was increasingly widely applied, and particularly to navigable rivers, as it was believed to ensure that the riparian states had equal navigation rights (Bouchez 1963). In stating that the thalweg would be determined by 'the main channel with the deepest draught and the easiest and clearest navigation',[11] the instructions received by the demarcators reflected the government's interest in ensuring these rights. This question was one of political significance for both countries: Waterways were not only valued as means of communication and territorial connection, but were also of military value, in the context of the arms race, as steamships and warships could navigate along them.

4.3 The Thalweg Method and the Distribution of the Islands in the Uruguay River

International law also established methods for determining the distribution of islands located in a frontier watercourse. The literature recognizes at least three ways of addressing the issue. The first method gives the line that defines the boundary along the river the power to distribute the islands between the riparian states. The second makes the allocation of the islands between states independent of the determination of the river border. Under the third, two separate lines have to be drawn: one to determine control of the river, the second to distribute control of the islands (Bouchez 1963, p. 815–817; Schröter 1992, p. 972–978; Caflisch 1989, p. 75–80).

[11] Convenio con el Brasil relativo a las instrucciones que servirá a la comisión mixta, que debe proceder a la demarcación de los límites entre ambos países, artículo 4.[Agreement with Brazil regarding the instructions that will be given to the joint commission, which is to demarcate the boundaries between the two countries, article 4]. (Argentina, Ministerio de Relaciones Exteriores y Culto 1910, vol. I, p. 524).

According to the instructions detailed in the 1898 agreements, the first of these methods was to be used: The thalweg line, in addition to being used to determine the boundary, was also to be used to differentiate the islands that would be under Argentine control from those that would be under Brazil's. The distribution would depend on each island's location relative to the bank and the thalweg. With regard to this, the agreement empowered the commissioners to propose any change they deemed fair to both countries.[12] According to Argentine Commissioner Ezcurra, studying the channels on the various islands in this part of the Uruguay River took up most of the time spent on work in the field:

> The differences in depth between them are so small that it requires a large number of soundings to establish the location of the thalweg and where the easiest channel for navigation lies. The consequence of this is that we need to measure appropriately spaced transversal sections of the river, which has forced us to measure nearly 250 kilometres of auxiliary lines—a much greater task than the triangulation itself (Argentina, Ministerio de Relaciones Exteriores y Culto. 1910, vol. 1, p. 532)

One of the difficulties they encountered in carrying out these soundings was the speed of the river currents. Ezcurra himself stressed the need for the soundings to be 'carried out under favourable conditions, on calm days so as to avoid little waves, and when the river is at its normal level, so as to avoid being carried off-course by currents'(Argentina, Ministerio de Relaciones Exteriores y Culto, 1910, vol. 1, p. 532). As the survey was carried out in increasingly rugged terrain, the waters became increasingly fast-flowing, which made it ever more difficult to take soundings and establish the thalweg. Even in sections of the river with gentle

[12] Tratado con el Brasil para completar por medio de acuerdo amigable y directo el establecimiento de la línea divisoria de los dos países, en parte determinada por el laudo arbitral del Presidente de los Estados Unidos de América, artículo 1. [Treaty with Brazil to complete, through a friendly and direct agreement, the establishment of the boundary line between the two countries, partly determined by the arbitral award of the President of the United States of America, article 1] (Argentina, Ministerio de Relaciones Exteriores y Culto 1910, vol. II, p. 137).

slopes, there were situations such as flooding that could make it impossible to carry out hydrographic work. Cerqueira confirmed this in a telegram that said '[The River] Uruguay is overflowing, islands submerged, soundings halted' (Argentina, Ministerio de Relaciones Exteriores y Culto 1910, vol. 1, p. 532).

The results of the sounding work and the establishment of the thalweg line are shown in the cartography relating to the Uruguay River (see Figs. 4.2 and 4.3), the Iguazu, and the islands surveyed. Wherever the demarcators decided to place a border marker on the mapped island, this was also included in the representation (see Figs. 4.3, 4.4 and 4.5).

Finally, the joint commission drew up a table (Table 4.1) that set out which islands were to be under Argentine control and which were to be included in Brazil's territory. This table, which was included in the 1910 Treaty of Rio de Janeiro, was organized following the course of the Uruguay River from south to north, then the Iguazu River, based on the order in which the demarcation work was carried out.

From Table 4.1, we can infer changes in control over the islands, associated with the thalweg changing how these were distributed between the nations along the riverbanks. These changes can be placed into four categories, as we describe below.

- Islands that had Portuguese place names (Grande de Saudade) and were transferred to Argentine control and islands that had Spanish place names (San Lucas Pequeña, del Roncador) and were transferred to Brazilian control. This leads us to infer that it was the thalweg line that caused changes in the ownership of the islands.
- Islands that had both Spanish and Portuguese names (Morcegos/Murciélagos, del Vado/do Vao, San Mateo/S. Matheus, San Lucas Pequeña or Isla del Cordeiro, Grande or Comprida/Larga, Burica/Mburica, Isla Grande/Ilha Grande, Saltiño/Saltinho). From this, we can infer that there had been some disagreement as to which of the two states owned them. Using the thalweg method would have helped to resolve conflicts of this kind by making it clear which country each one belonged to.
- Archipelagos whose islands were divided between the two states through the use of the thalweg definition: San Lucas Grande (Argentine) and San Lucas Pequeña (Brazilian), the latter located to the north of San Lucas Grande; San Isidro (a group of islands made up of three larger islands and several smaller ones, all of which remained under Brazilian control, except for one larger islet that remained under Argentine control).
- Archipelagos with place names in both Portuguese and Spanish that were divided between the two states through the use of the thalweg definition: Piratinin/Piratini (composed of four islands and an islet, of which the three easternmost islands and the islet belong to Brazil, while the western island belongs to Argentina); Sam Javier/San Javier (three islets belong to Brazil, while the islet to the north of the three Brazilian ones remains under Argentine control); Santa Lucía Inferior or Oriental/Santa Luzía inferior (under Brazilian control) and Santa Lucía Superior or occidental or Santa Luzía Superior (under Argentine rule); Sam Javier (three islets included in the territory of Brazil) and San Javier (included in the territory of Argentina); San Lucas Grande (Argentina) and San Lucas Pequeña/Isla del Cordeiro (Brazil). In these cases, establishing the thalweg line not only implied resolving disagreements over islands whose sovereignty was disputed by Argentina and Brazil, it also implied dividing archipelagos.

4.4 Demarcation Work on the Iguazu Falls

The demarcators consistently used the thalweg method to determine the border along the Iguazu River. However, the cataracts along the river made it difficult for the demarcators to establish geographical locations from their boats. This

Fig. 4.2 Map of the border between the Argentine Republic and the United States of Brazil (1900–1904). Scale: 1:1,000,000. Topographical maps of the Islands in the Uruguay River and the Islands in the Iguazu River. *Source* Argentina, Ministerio de Relaciones Exteriores y Culto 1910, vol. III

Fig. 4.3 Map known as 'Carta de Parte da Fronteira entre a República dos Estados Unidos do Brasil e a República Argentina.' Scale: 1:500,000. The cartography represents the islands surveyed and the border markers (in red). *Source* BNB

led the demarcators to use practices, techniques and instruments that had not been used previously (Capilé and Vergara 2014). The Brazilian delegation first carried out research to find out how the boundary definition work between the United States and Canada had been completed at Niagara Falls.[13] The Brazilian Commissioner Cerqueira then requested that a phototheodolite be sent to the joint boundary commission, as he believed that this would enable them to carry out topographic work on inaccessible and rugged terrain, over which large volumes of water flowed and which was prone to flooding. The Laussedat phototheodolite made it possible to take photographs of the terrain in order to obtain topographical maps.

The joint commission's efforts resulted in the maps named 'Cataractas do Iguaçu ou Salto de S. Maria' (Brazilian version) and 'Cataratas del Iguazú' (Argentine version), both at a scale of 1:5000. They contain more than 60 photogrammetric points highlighted in red (Capilé and Vergara 2014, p. 111). Both the Brazilian and the Argentine maps contain details of the topography and vegetation and are accompanied by a set of photographs that surround the map itself. The Brazilian map has sixteen photographs, while the Argentine map has seven (see Figs. 4.6 and 4.7).

[13] Cerqueira wrote to the Brazilian Minister of Foreign Affairs: 'It is impossible to take soundings in the region of the above-mentioned [Great Iguazu] Falls to establish the thalweg, and, as it seems appropriate to me to study this matter immediately, I request Your Excellency to obtain, through our legation in the United States of America, the volume of Supreme Court judgments containing the decision in the case (analogous to the Iguazu Falls case) of the demarcation of Niagara Falls with Canada.' AMREBI. Comissão de Limites com a República Argentina. File without number, May 27 1902 (Box 434, pack 3).

Fig. 4.4 Detail of the map known as 'Carta Parte da Fronteira entre a Republica dos Estados Unidos do Brasil e a República Argentina.' Scale: 1:500,000. This shows the Isla Grande, where a border marker was erected. *Source* BNB

We believe these photographs helped to place the landscape of the falls within the national representations of each country (Zusman 2022). However, those seen on the Argentine map specifically seek to focus attention on the geoforms that are part of the border. In other words, as Mazzitelli Mastricchio (2012) argues, they are a visual proof of the measurements taken in the field. These photographs show different panoramas of Salto Unión (the waterfall where the thalweg is located), of the waterfalls that, after the demarcation, came to belong to Argentina and Brazil, and of the landscape that the boundary was acquiring along the lower Iguazu River. The co-presence of a photograph and a map creates a dialogue between these two objects. Thus, following Mazzitelli Mastricchio (2012), while the photograph helped viewers to imagine the border drawn in the cartographic representation, the cartographic image helped viewers to imagine the border in the photograph.

4.5 Geographical Coordinates and the Survey of Boundary Markers

Taking geographical coordinates was part of the process of determining the locations where markers denoting the international boundary would be placed. The joint demarcation commission gave a high priority to taking coordinates at the headwaters of the Pepiri-Guazu and San Antonio rivers, on the islands, and at the key

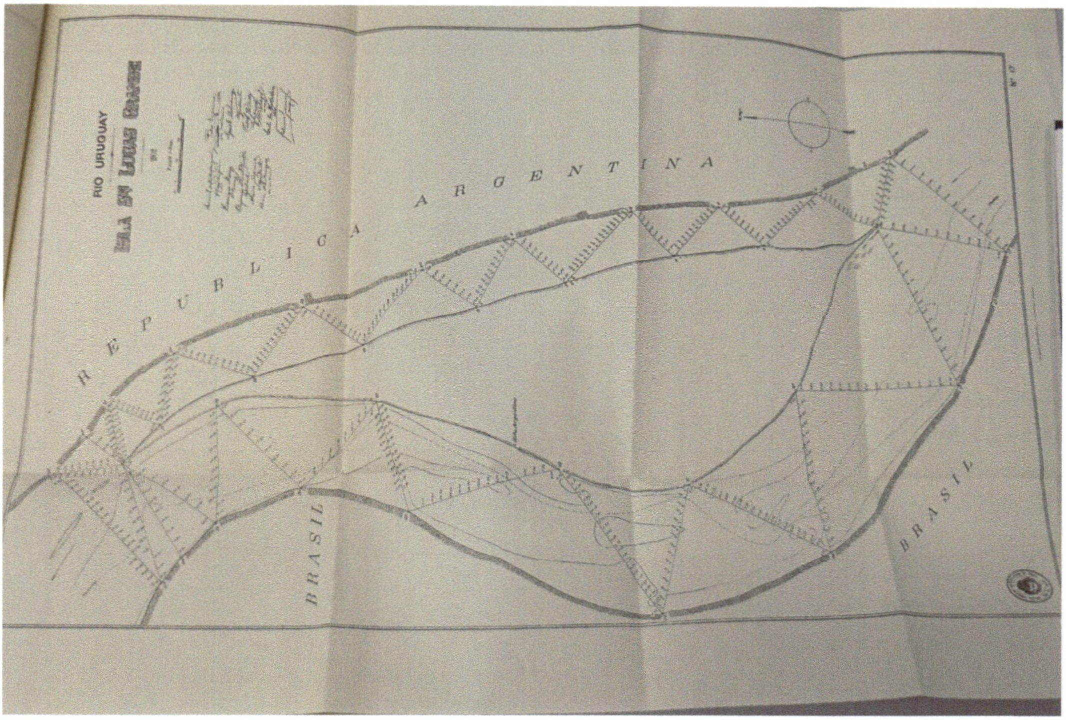

Fig. 4.5 Isla Grande under Argentine control. The representation makes it possible to identify the sounding work carried out to determine the main channel. In this case, the main channel is shown to run close to the Brazilian bank, leaving Isla Grande under Argentine control. *Source* BNB

confluences in the delimitation process: that of the Cuareim River with the Uruguay, that of the Uruguay River with the Pepiri-Guazu, that of the San Antonio with the Iguazu, and that of the Iguazu with the Paraná.

This task was far from easy to carry out. The geographical coordinates were taken by observing stars and other celestial bodies, using precision instruments. Taking these measurements could 'take days or even weeks, depending on the atmospheric conditions'.[14] Positional astronomy was carried out at night if possible, as this ensured greater accuracy. In the course of a night of observation, the astronomers would measure the same star at different times. On some occasions, instruments and stars might both be used alternately, in order to calculate an average of the results across both methods and thus minimize the potential measurement errors at a given location (Capilé and Vergara 2013).

To take the coordinates, the commissions used a range of instruments, including sextants, theodolites and portable chronometers. After the turn of the twentieth century, with improved chronometers and the spread of telegraph networks, measuring longitudes became less dependent on astronomy, as they could be obtained by comparing the local time with a time referenced at a fixed place equipped with an astronomical observatory. An official letter from Brazilian commissioner Cerqueira, dated 14 March 1903 and announcing that the demarcation work had been completed, thus states: 'in places where there was a telegraph, the longitudes were referenced to the observatories of either Rio de Janeiro or Córdoba, depending on whether it was on Brazilian or Argentine

[14] AMREBI. Comissão de Limites com a República Argentina. File n. 23, April 10 1902 (Box 434, bundle 3).

Table 4.1 Distribution of islands and islets in the Uruguay River, from the height of the Cuareim confluence to the mouth of the Pepiri-Guazu, and from the confluence of the San Antonio and Iguazu Rivers towards the confluence of the Iguazu with the Paraná

Argentina	Brazil
Pacú (island)	
Grande de Saudade (island)	
	Yapeyú/Japejú (island)
	De la Cruz (island and islet)
Chaparro (island)	
Aguapey (island)	
	Palomas (island)
Murciélagos/Morcegos (island)	
Tacuaras Inferior	
	Cuadrada (island)
	Butuhy Grande and Butuhy Chica (islands)
Del Vado/do Vao (island)	
Islote de Tigre (island)	
Isla del Cuai (island)	
Isla de Santa Ana (island)	
	Santa Lucía Inferior or Oriental/Santa Luzía inferior (island)
Santa Lucía Superior or Occidental/Santa Luzía Superior (island)	
De Vargas (island)	
San Mateo/S. Matheus (island)	
Sarandi (group of four islets)	
	Mercedes (one island and four islets)
	Pequeña (island)
Grande (island)	
	De Los Garruchos (group of four islets)
San Lucas Grande (island)	
	San Lucas Pequeña/ Isla del Cordeiro (island and islets)
	Tacuaras Superior or de Arriba (islands of the and neighbouring islets)
Cerrito (island)	
	Piratinin/Piratiny (four islands and one islet)
Piratini (one of the four islets of that name)	
	San Isidro (group of islets, including three larger ones and several small ones)
San Isidro (one of the three larger islets in the group of this name)	
	Santa María (group of three larger islands and several islets)
Ijuhi (two islets)	

(continued)

Table 4.1 (continued)

Argentina	Brazil
Itacaruaré Chica (island)	
	Itacaruaré Grande (island and its group)
	Sam Javier/San Javier (three islets)
San Javier (islet above the three Brazilian islets of San Javier)	
	Cumanday (group of three islands and several islets)
	Grande or Comprida/Larga (island)
Chico Alférez (group of islets and reefs)	
	Del Roncador (reefs)
	Del Bugre or de los Bugres (island and three islets)
Del Borracho (4 islets)	
	Biguá (island)
Canal Tuerto/Canal Torto (small islets)	
	Jacaré/Yacaré (island)
	Saltinho/Saltiño (islets and reefs)
Chafariz (two islands and several islets)	
	Burica/Mburica (two small islands and several islets)
Pucha Para Atraz (islet)	
Dino or Nao (island)	
Pepiry Guazú/Pepiry Guassú (islet)	
	Pesqueiro (island)
Isla Grande/ Ilha Grande (group of four islands and an islet)	
	Islote de las Tacuaras
Isla de San Agustín or Sola	
Islands, islets and reefs preceding the Iguazu Falls and adjacent to the left (Argentine) bank	Islands, islets and reefs preceding the Iguazu Falls and adjacent to the right (Brazilian) bank

Source Created by the authors based on the contents of article 3 of the 1910 Treaty of Rio de Janeiro, Argentina, Ministerio de Relaciones Exteriores y Culto (1910), vol, II, pp. 287–304. Names given to the islands, islets and archipelagos in the original document were respected

territory.'[15] In the same letter, Cerqueira also states that 'the points where there was no telegraph were connected to the nearest points with telegraph services by means of light signals at night'.[16]

Light signals, or rather fire signals, were transmitted by producing sparks through explosions of gunpowder carried out at 'previously agreed intervals at an intermediate point between a location whose longitude was already known and another whose longitude needed to be established' (Capilé and Vergara 2013, p. 261). This method required the time at each point to be carefully established in advance, as it was the difference between the times when the explosion occurred that would allow the difference in longitude to be calculated. It was used mainly on the high ground where the headwaters of the Pepiri-Guazu and San Antonio rivers were located.

[15] AMREBI. Comissão de Limites com a República Argentina. File n. 41, March 14 1903 (Box, 434, bundle 3).

[16] AMREBI. Comissão de Limites com a República Argentina. File n. 41, March 14 1903 (Box 434, bundle 3).

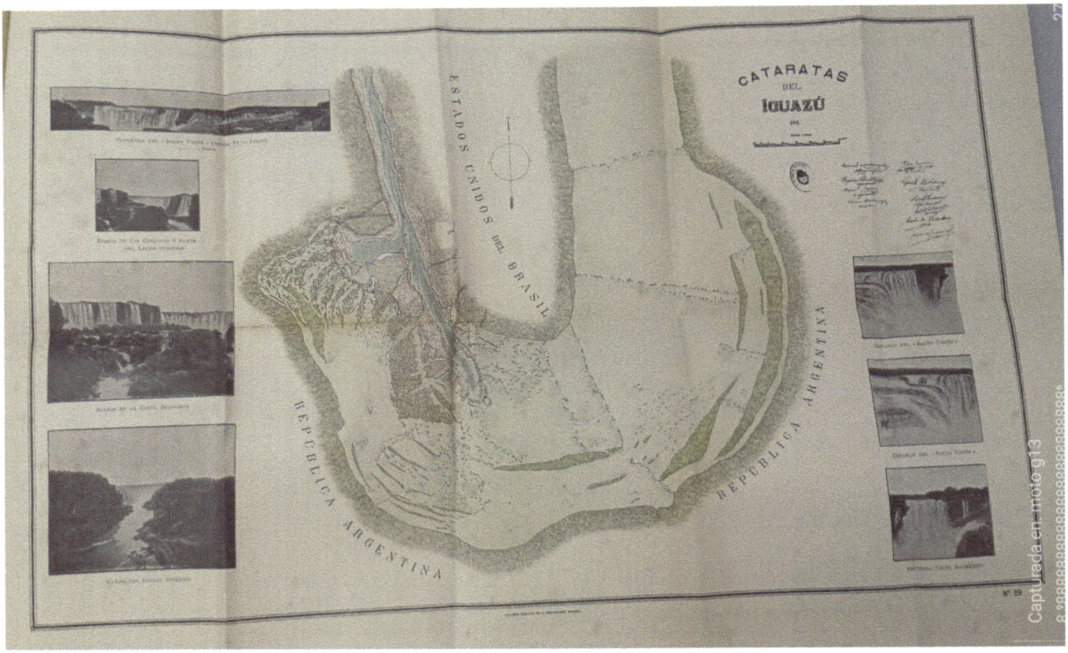

Fig. 4.6 'Cataratas del Iguazú' map, scale 1:5000. Topographical maps of the islands in the Uruguay River and the islands in the Iguazu River Argentina. *Source* Argentina, Ministerio de Relaciones Exteriores y Culto (1910), vol III

As with the telegraph and fire signals, the use of chronometers followed the principle of determining longitudes by comparing the local times at the desired points with the time at an observatory, where they were set. At first glance, the chronometers did not have the same disadvantages as fire signals, as they were not affected by atmospheric conditions. However, transporting the chronometers, particularly along fast-flowing waterways such as the Uruguay River, caused their mechanisms to become deregulated and affected their operation.[17] There is reason to believe that lessons were learnt and that improvements were made to the conditions in which the chronometers were transported, as the joint commission, meeting at its fourth conference on 24 January 1903, decided that the longitudes between the mouth of the Iguazu and the ports of Piraí (Brazil) and Posadas (Argentina) would be determined by the use of chronometers (Argentina, Ministerio de Relaciones Exteriores y Culto 1910, vol. II, p. 248).

Some of the coordinates determined the placement of the border markers. In total, 95 markers were placed. These include the first two (one on the Argentine side and the other on the Brazilian side), erected where the demarcation began (at the confluence of the Cuareim and Uruguay rivers), the 35 erected on the islands (seventeen on Argentine islands and eighteen on Brazilian islands), the 50 obelisks shared by both countries (two primary markers, three secondary, and forty-five tertiary) located between the headwaters of the Pepiri-Guazu River and those of the San Antonio (watershed ridges dividing the waters of the tributaries of the Paraná River from those draining into the San Antonio and Iguazu rivers), and the two last markers, one on the Argentine side and one on the Brazilian side, denoting the confluence of the Iguazu and Paraná rivers. The boundary markers were made by masons and stonemasons, out of materials obtained in the locations where they were erected. The first Argentine

[17] On the transport and use of instruments in exploration, see: Macdonald and Withers (2016), Bourguet et al. (2002), Collini and Vannoni (2005).

Fig. 4.7 Detail of the 'Cataratas del Iguazú' map showing the location where the phototheodolite was placed. Topographical maps of the islands in the Uruguay River and the islands in the Iguazu River. *Source* Argentina, Ministerio de Relaciones Exteriores y Culto (1910), vol. III

border marker, which was located 'on the right bank of the Uruguay River, on a high ravine, opposite the confluence with the Cuareim River,' was thus built out of 'brick masonry, sand and Portland cement' (Argentina, Ministerio de Relaciones Exteriores y Culto, 1910, vol. II, p. 530), while the Brazilian one, located on the opposite bank of the Uruguay River, very close to the confluence of the Cuareim, was made of stone.

These boundary markers gave the border a material form on the ground (Fig. 4.8), but they would also be represented in the cartography. This is demonstrated by the map entitled 'Situação topográphica dos marcos principais' (Fig. 4.9), which shows the four confluences mentioned above: the Cuareim River with the Uruguay River, the Uruguay River with the Pepiri-Guazu, the San Antonio with the Iguazu, and the Iguazu with the Paraná. The dry border area corresponding to the watershed between the Pepiri-Guazu and San Antonio rivers is also exhibited. The boundary markers indicated on the maps of the river confluences are in dialogue with the sketches of obelisks included in the same maps.

Thus, the maps represented the markers located in Brazil (Marco Brasileiro das Barras do Quarahim, Iguassú, do Pepiry Guassu, do Santo Antonio[18]) and on the headwaters of the Pepiri-Guazu and San Antonio. Unlike with other maps produced by the demarcation commissions that were intended to show the demarcation line, here the aim was to register on the map itself the devices that gave a material form to the border on the ground. In this case, the cartography helps the viewer to imagine the landscapes resulting from the incorporation of the markers on the ground, thus providing another way of representing the border.

---

[18] Names given to the marks in the original map were respected.

Fig. 4.8 Brazilian marker at the confluence of the Iguazu River with the Paraná. *Source* AMREBI, Comissão de Limites com a República Argentina. Book 5. (1905)

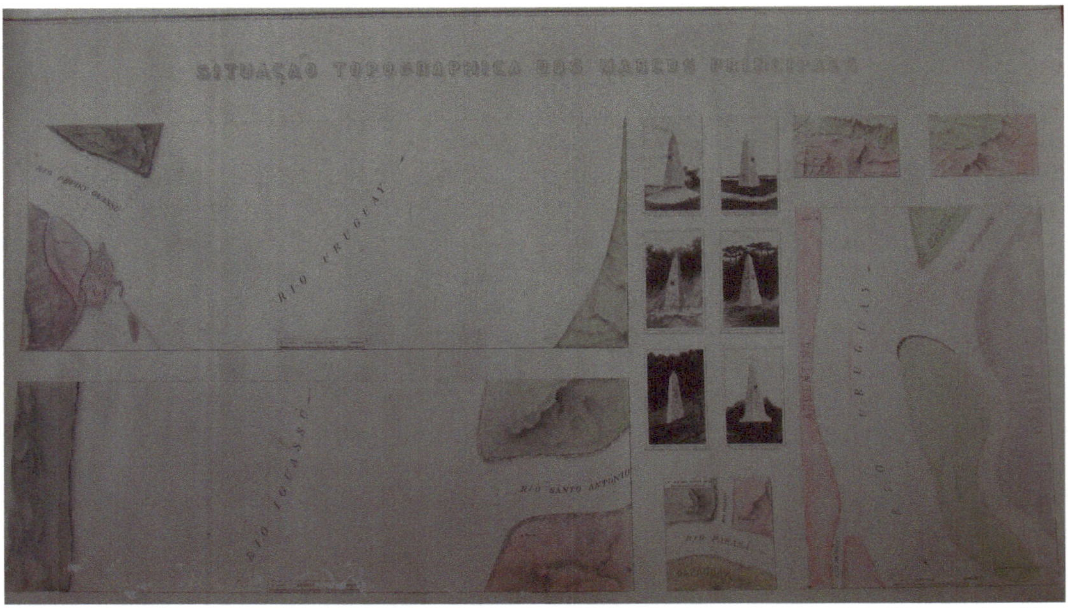

Fig. 4.9 Topographical situation of the main boundary markers. *Source* AMREBI, Comissão de Limites com a República Argentina. Book 5. (1905)

4.6 Conclusion

In this chapter, we have described the process by which the border between Argentina and Brazil was established. Through the use of sources in repositories belonging to both these countries, we have been able to observe how the distribution strategies agreed upon by the members of the joint demarcation commission were implemented. These strategies involved practices, techniques and instruments for overcoming the difficulties encountered in the field. Determining the thalweg in the Uruguay and Iguazu rivers first enabled the commission to distinguish the islands that were to be controlled by Argentina from those to be controlled by Brazil, and, second, enabled it to identify which waterfalls should be part of the territory of each state. The scientific and technical method of using the thalweg prevailed over other methods for defining borders along watercourses and toponymic or historical approaches to allocating islands. At the same time, the use of the phototheodolite made it possible to overcome the difficulties of demarcating the border in the area of the falls. Finally, practices, techniques and instruments that made it possible to overcome the difficulties of calculating coordinates determined the locations where markers should be placed, particularly in areas with headwaters, river confluences or islands.

According to some authors, these border markers were not erected solely to demarcate territories, but also to celebrate peace and the resolution of the dispute (Louvain 2018). The completion of the demarcation process did indeed have diplomatic implications. Both Argentina and Brazil showed a desire to strengthen ties of friendship and cooperation. As a demonstration of this, after the protocol ratifying the arbitration award by the president of the United States of America was signed on 6 October 1898, the president of Argentina, Julio Roca, visited Brazil. Besides being the first visit ever made to Brazil by a foreign president, Roca's visit became a symbolic ritual that was repeated whenever the two nations wanted to create closer ties. The president of Brazil, Manuel de Campos Salles, reciprocated by visiting Argentina in 1900 (Candeas 2005; Doratioto 2014). Furthermore, the political classes in both countries recognized the negative effects that the prospect of armed conflict could have on the agro-export economic models that were becoming established.[19] This chapter thus not only aims to make a contribution on graphic representations of boundary definition processes along watercourses, but also to reflect on the links between demarcation activities and diplomatic activities in countries that were establishing how they would be included in the world system.

Archival Sources

AMREBI: Arquivo do Ministério das Relações Exteriores do Brasil - Itamaraty (Archive of the Brazilian Ministry of Foreign Affairs—Itamaraty], Rio de Janeiro.

BNB: Biblioteca Nacional do Brasil, Rio de Janeiro.

[19] During this period, Argentina and Brazil were becoming part of the capitalist economy as a result of their exports of raw materials. Argentina was selling cereals and cattle to Europe, while Brazil was experiencing an economic boom driven by Amazonian rubber. At the same time, a regional circuit for the flow of specific goods was identified, contributing to this economic growth. Argentina supplied food to Manaus and Belém in northern Brazil. Both countries were deeply connected by the production and trade of yerba mate. Argentina was the main consumer of yerba mate produced in Brazil, often in settlements held by Argentines. In a letter dated 20 April 1904, Cerqueira emphasized the importance of ending the commission's work, citing the need to incorporate the area into yerba mate production. This work was necessary, he wrote, 'due to the importance of the border and the demands of the terrain, in which the thalwegs of the two great rivers had to be determined in order to allocate a large number of islands and carry out the meticulous demarcation of a watershed in a region of great importance today for the cultivation of yerba mate' AMREBI. Comissão de Limites com a República Argentina. File without number. (Box 434, bundle 3).

References

Altic M (2023) From borderlands to boundary lines: mapping on the edges of the Ottoman empire. Imago Mundi 74(2):189–216. https://doi.org/10.1080/03085694.2022.2130521

Argentina Ministerio de Relaciones Exteriores y Culto (1910). La frontera argentino brasileña. Estudios y demarcación general (1887–1904), 3 vols. Talleres Gráficos de la Penitenciaría Nacional, Buenos Aires

Bouchez LJ (1963) The fixing of boundaries in international boundaries rivers. Int Comp Law Q 12(3):789–817. https://doi.org/10.1093/iclqaj/12.3.789

Bourguet MN, Licoppe C, Sibum HO (2002) Instruments, travel and science: itineraries of precision from the seventeenth to the twentieth century. Routledge, London

Caflisch L (1989) Les frontières interétatiques se rattachant aux cours d'eau internationaux. Recueil des Cours De L'académie De Droit International De La Haye 219(7):62–103

Candeas AW (2005) Relações Brasil-Argentina: uma análise dos avanços e recuos. Revista Brasileira De Política Internacional 48(1):78–213. https://doi.org/10.1590/S0034-73292005000100007

Capilé B, Vergara M (2013) Astronomia e território: a Comissão Demarcadora de Limites entre Brasil e Argentina. In: Matsuura O (ed) História da Astronomia no Brasil. CEPE, Recife, pp 329–350

Capilé B, Vergara M (2014) O desenvolvimento do fototeodolito e seu uso na fronteira entre Brasil e Argentina História Unisinos. História Unisinos 18(1):104–113. https://doi.org/10.4013/htu.2014.181.10

Cervo A, Bueno C (1992) História da política exterior do Brasil. Ática, São Paulo

Chianelli TD (1989) Cien años de panamericanismo. Todo Es Historia 270:7–34

Collini S, Vannoni A (2005) Les Instructions Scientifiques pour les Voyageurs (XVIIe-XIXe Siècle). L'Harmattan, Paris

Cutolo VO (1969) Nuevo diccionario biográfico argentino vol 2. Elche, Buenos Aires

Di Fiore L (2017) The production of borders in nineteenth–century Europe: between institutional boundaries and transnational practices of space. Eur Rev History—Revue Européene d' Histoire 24(1):36–57. https://doi.org/10.1080/13507486.2016.1183594

Donaldson JW (2002) Politics and scale in boundary-making: the work on the boundary commissions. J Hist Geogr 34(3):393–396. https://doi.org/10.1016/j.jhg.2008.02.001

Doratioto F (2014) O Brasil no Rio da Prata. Fundação Alexandre de Gusmão, Brasília

Erbig, J Jr (2020) Where Caciques and Mapmakers meet. Borders making in Eighteen South America. University of North Carolina Press, Chapel Hill

García-Álvarez J, Puente-Lozano P (2025) 'Borderism': imaginative geographies and the production of modern boundaries in Spain and Portugal, 1840–1870. Ann Am Assoc Geogr 115(1):131–147. https://doi.org/10.1080/24694452.2024.2402030

García-Alvarez J, Puente-Lozano P (eds) (2022) Beneath the lines. Borders and boundary-making from the 18th to the 20th century. Springer, Cham

Garay C (2012) Las carreras armamentistas navales entre Argentina, Chile y Brasil (1891–1923). Historia Crítica 48(1):39–57. https://doi.org/10.7440/histcrit48.2012.03

Garay Vera C (2021) El equilibrio de poder como debate en las relaciones internacionales del Cono Sur Americano (1830–1910). Historia 396 11(2):199–230. https://historia396.cl/index.php/historia396/article/view/546/222

Herzog T (2015) Frontiers of possession: Spain and Portugal in Europe and the Americas. Harvard University Press, Cambridge

Jiménez Cabrera D, Garay Vera C (2021) El equilibrio de poder como debate en las relaciones internacionales del Cono Sur Americano (1830–1910). Historia 396 11(2):199–230. https://historia396.cl/index.php/historia396/article/view/546/222

Lacoste P (2001) Chile y Argentina al borde de la guerra (1881–1902). Anuario del Centro de Estudios Históricos Profesor Carlos S. A. Segreti 1:301–328. https://doi.org/10.52885/2683-9164.v.n1.23316

Lois C (2019) De las fronteras coloniales del imperio hispánico en América a los límites internacionales entre Estados latinoamericanos independientes: génesis de la imposibilidad de un mapa político de Sudamérica consensuado. Revista de Historiografía 30:207–222. https://doi.org/10.20318/revhisto.2019.4749

Louvain P (2018) O Marco das Três Fronteiras: Símbolo da integração latino-americana e de resolução pacífica dos litígios internacionais. Anais do XVII Encontro Estadual de Histórica da ANPUH-SC, Joinville. https://www.encontro2018.sc.anpuh.org/resources/anais/8/1538362444_ARQUIVO_LOUVAIN.OMarcodasTresFronteiras.ANPUH-SC.pdf

Magnoli D. (1997) O corpo da pátria: imaginação geográfica e política externa no Brasil (1808–1912). Ed. UNESP, São Paulo

Macdonald F, Withers CWJ (2016) Geography, technology and instruments of exploration. Routledge, London

Martins Filho JR (2023) A marinha brasileira na era dos encouraçados, 1895–1910. Fundação Getúlio Vargas, Rio de Janeiro

Mazzitelli Mastricchio M (2012) El aspecto sensible de las prácticas cartográficas: el uso de la fotografía en los trabajos topográficos de la Dirección de Minas, Geología e Hidrología (1940) In: Lois C, Hollman V (eds) Geografia y cultura visual. Los usos de las imágenes en las reflexiones sobre el espacio. Prohistoria Ediciones-Universidad Nacional de Rosario, Rosario, pp 329–350

Matus Matus MI, Aranda Bustamante G (2016) A 100 años del ABC: desafíos y proyecciones en el marco de la integración regional. CESIM-IEI, Santiago de Chile

Mc Gann TF (1957) Argentina, United States and the inter-American system (1880–1914). Harvard University Press, Cambridge (USA)
Morgenfeld LA (2011) Vecinos en conflicto. Argentina y Estados Unidos en las Conferencias Panamericanas (1880–1955). Continente, Buenos Aires
Moniz Bandeira LA (2004) Argentina, Brasil y Estados Unidos. Norma, Buenos Aires
Moniz Bandeira LA (2010) Presencia de Estados Unidos en Brasil. Dos siglos de Historia. Corregidor, Buenos Aires
Nascimento JO (2015) Cerqueira, Dionísio. In: Abreu A (ed) Dicionário Histórico Geográfico da Primeira República (1889–1930). Editora Fundação Getúlio Vargas-CPDOC, Rio de Janeiro: 2326–2334 (e-book)
Oszlak O (2004) La formación del Estado Argentino. Ariel, Buenos Aires
Sahlins P (1990) Natural frontiers revisited: France's boundaries since the seventeenth century. Am Hist Rev 95(5):1423–1451. https://doi.org/10.2307/2162692
dos Santos RSB (1977) Hidrografia. In: Galvão MV (ed) Geografia do Brasil. Região Sul vol. 5:110–114. IBGE, Rio de Janeiro
Schröter F (1992) Les frontières interétatiques. Annuaire Française De Droit International 38:948–982. https://doi.org/10.3406/afdi.1992.3103
Sotomayor TM (1996) Estados Unidos y el panamericanismo: el caso de la I Conferencia Internacional Americana (1889–1890). Hist Mex 45(4):759–781
Strandsberg J (2017) Cartopolitics, Geopolitics and Boundaries in the Arctic. Geopolitics 17(4):818–842. https://doi.org/10.1080/14650045.2012.660581
Tamayo Pérez LMO (2022) Mexican geographical engineers and demarcation of the borders of Mexico. In: García-Álvarez J, Puente-Lozano P (eds) Beneath the lines borders and boundary-making from the 18th to the 20th century. Springer, Cham, pp 73–89
Urteaga L (2022). The demarcation of the border of the Spanish Protectorate in Morocco (1913–1931): Chronicle of a failure. In: García-Álvarez J, Puente-Lozano P (eds), Beneath the lines. Borders and boundary-making from the 18th to the 20th century. Springer, Cham, pp 139–158
Ventura de Marco R (2023) Percepción de amenazas y estrategias de equilibrio: Argentina y la carrera armamentista naval del Cono Sur (1906–1914). Estudios Internacionales 55(205):117–138. https://doi.org/10.5354/0719-3769.2023.70305
Wallerstein I (2004) World system analysis. Duke University Press, Durham and London
Wimmer A, Glick Schiller N (2002) Methodological nationalism and beyond: nation-state building, migration and the social sciences. Global Networks J Transnat Affairs 2(1):301–334. https://doi.org/10.1111/1471-0374.00043
Zusman P (2017) La técnica y la definición de fronteras. Revista De Geografía Norte Grande 66:49–60. https://doi.org/10.4067/S0718-34022017000100004
Zusman P (2022) La construcción de las Cataratas del Iguazú como paisaje argentino y su incorporación en la escena global. Treballs de la Societat Catalana de Geografía 93:153–173. https://doi.org/10.2436/20.3002.01.223
Zusman P, Hevilla C (2014) Panamericanismo y arbitraje en conflictos de límites. La participación de Estados Unidos en la definición de la frontera argentino-chilena en la Puna de Atacama (1899). Cuadernos de Geografía 23(2):95–106
Zusman P, Nunes-Pereira S (2022) Boundary commissions under the Treaty of Madrid and their role in the reinvention of the Rio de la Plata frontier (1752–1759). In: García-Álvarez J, Puente-Lozano (eds) Beneath the lines. Borders and boundary-making from the 18th to the 20th century. Springer, Cham, pp 93–108
Zusman P, Nunes Pereira S, Moraes C (2023) Circulaciones fronterizas. Políticas de colonización en torno al área de conflicto Palmas/Misiones (1850–1930). In: García-Hernán D, García-Álvarez J, Curnis M (eds) Fronteras del Mundo Hispánico: ideas, conflictos, identidades. Sílex, Madrid, pp 253–275

5 Drawing Lines in the Sand: Cartography and Boundary-Making in Colonial Western Sahara

Ana Raya-Collado and Jacobo García-Álvarez

Abstract

This article aims to reflect on the role of cartography in the delimitation and demarcation of colonial spaces associated with desert environments. The first part of the chapter reflects on the particularities of boundary-making processes in such environments, with special attention to the frequent use of geometric boundaries and to the role of maps in the appropriation and colonization of these spaces, as well as in the boundary disputes linked to them. As a case study, it will analyse the example of Western or Spanish Sahara, whose borders were defined in the Spanish-French conventions of 1900, 1904 and 1912, but were neither delimited nor demarcated until the 1950s, when a Joint International Commission was created to carry out these tasks. As the chapter argues, the cartography produced during this period was key in the process of geometrization and administrative purification of the borders of this mostly desert region, facilitating the transition from a zonal frontier, such as the one that separated the Spanish and French colonial administrations for several decades, to a linear boundary, linked to the European modern ideas of sovereignty and territoriality.

Keywords

Cartography · Boundary-making · Deserts · Geometric boundaries · Colonialism · Partition of Africa · Western Sahara · Spain · France · Iyil salt pans

5.1 Introduction

Thus, we find in some of the remoter portions of the globe long international boundaries defined by straight lines. Now Nature abhors a straight line, and Nature will have no hand in presenting opportunity for a useful and easily demarcated boundary, that is absolutely straight. Nothing but the necessity imposed by ignorance can justify the adoption of the straight line. It is inelastic, allowing of none of that give and take in the business of mutual settlement on the field which is an essential factor in a rapid and reasonable agreement. There are innumerable features, small in themselves maybe, but locally important, that may require a divergence from a rigid line. There may be the question of water supply (or perhaps of oil) to consider, and wells may be unevenly distributed. Or, again, a village may become separated from the tribal community to which it rightly belongs. Indeed, the

This work has been supported by the Spanish State Research Agency under Grant Number PID2020-114088GB-I00/AEI/10.13039/501100011033, and by the Ministry of Science, Innovation and Universities under the FPU Grant Number 21/02850.

A. Raya-Collado (✉) · J. García-Álvarez
Department of Humanities: History, Geography and Art, Universidad Carlos III de Madrid, Getafe, Madrid, Spain
e-mail: araya@hum.uc3m.es

J. García-Álvarez
e-mail: jacobo.garcia@uc3m.es

J. García-Álvarez and A. Reyes Novaes (eds.), *Mapping the Contested Line*, Historical Geography and Geosciences, https://doi.org/10.1007/978-3-032-29889-8_5

complications arising from a straight line definition are almost infinite, and they all add to the difficulties of demarcation and tend to foster ill-feeling and resentment, not only between the representatives or commissions of the two countries engaged in setting up a hedge between their respective interests, but among the people themselves, whose interests are often the last to be considered (...) It is usually where the expansion of white races has led to international complications caused by the process of supporting claims in regions only half explored and possibly represented by an absolute blank in the world's map, and where the division between international interests appeared to require immediate settlement without the delay of scientific exploration, that resort has been made to such a weak substitute for rational and scientific frontier definition. (Holdich 1916, pp. 169–171)

Speaking very generally, the most suitable boundaries in deserts may be arbitrary lines, perhaps determined on the ground by the demarcation commission. The low value of the land, in most cases, permits the granting of much power to the demarcators. Attempts to pick naturally marked lines in deserts may only lead to trouble. Mountains in deserts frequently serve as foci of routes or centres of population because of the water supplies they afford and hence may be bad boundary features (...). The temptation to delimit boundaries along the watercourses indicated on maps may lead to difficulties in demarcation (...). Because of the sparse population, the paucity of naturally marked sites, and, in many cases, the rudimentary state of geographical information on deserts, geometrical boundaries are common in them. It may be said here that, if a geometrical line is drawn across a desert, very large powers to deviate from it should be given the demarcation commission. These powers are needed even if the desert is completely uninhabited, for the delimited line may be very expensive to demarcate and, in fact, may be inaccessible except at considerable risk of life (...) If the desert has inhabitants, settled or nomadic, a geometrical line is almost certain to be capable of improvement (...) Since water is the most important resource in most deserts, boundary-making should give it great consideration. Oases should not be cut from their water sources, which might happen if a geometrical boundary is followed slavishly. (Jones 1945, pp. 105–107)

The two extensive quotations excerpted above, taken from two of the major classic authors in the literature on boundary-making (the English Thomas Holdich and the American Stephen Jones), summarize quite well the main virtues and problems associated with the delimitation of boundaries using straight lines in general, and in desert environments in particular. For both geographers, the drawing of straight borders, whether astronomical or geometric, can be useful especially as a system for allocating, from the comfort of diplomatic offices and conference halls, a geographically unknown and cartographically virgin territory, such as many of the great deserts of the interior of Africa and Asia were at the time of their colonial partition. However, the demarcation of such borders on the ground involved multiple difficulties, such as the economic and human costs of surveying and erecting boundary markers in a shifting, uniform and climatically extreme landscape, or those arising from dividing between countries certain spaces and natural resources (especially those linked to the presence of water) that were vital to the way of life of local populations, frequently nomadic. It is for these and other important reasons that the seemingly simple task of drawing lines in the desert sand has often become a complex and controversial challenge.

This contribution examines a case that is emblematic in many ways of the relationship between maps, border delimitation and European occupation of colonial desert spaces: that of colonial Western Sahara, also known as Spanish Sahara or Spanish Western Sahara (1884–1975), a territory of approximately 270,000 square kilometres largely occupied by rocky and sandy deserts. The borders of this territory were defined in the Spanish-French Conventions of 1900, 1904 and 1912 but were neither delimited nor demarcated until the 1950s, when a Joint International Commission was created to carry out these tasks. As the chapter argues, the cartography produced during this period was key in the process of geometrization and administrative purification of the borders of this mostly desert region, facilitating the superimposition of the colonial powers' own territoriality—linked to the linear border, state organization and exclusive sovereignty—to the detriment of the zonal, fluid and imprecise character that the movements of the native nomadic tribes imprinted on the borders. But maps were also key to other dimensions of the process of colonial legitimization and appropriation of this territory, whose border delimitation, initially defined on essentially geometric criteria, led to various disputes—for the most part cartographic—between France and Spain.

The precolonial and colonial cartography of Spanish Sahara has been the subject of a number of previous studies that have described, among other aspects, some of the main cartographic images of the region as well as the special characteristics of mapping processes in this type of geographic environment, including, in some cases, the involvement of the native populations in such processes (Martín Escorza 2008; Rodríguez Esteban 2010). Particular attention has been given to the pioneering work carried out by the Spanish scientific-commercial expedition of 1886 (González Bueno 2008); and especially the 1:500,000 scale map of 1943–1949 by the Geographic Service of the Spanish Army, which was studied in depth by José Antonio Rodríguez Esteban (Rodríguez Esteban 2011). Much less attention has been devoted to other key episodes directly related to the delimitation of colonial borders in the region, which will be examined here, such as the cartographic disputes between the Spanish and French authorities in the 1930s and 1940s over the location of boundary wells near the Iyil salt pans, or the work of the French-Spanish boundary commissions which, in the 1950s, undertook the first effective modern demarcation of borders.[1]

The chapter is structured in two parts, in addition to this introduction and the conclusion. The first part will address general issues on the relationship between maps and the defining of borders in colonial desert territories. As argued in this section, the role of maps in boundary-making processes in colonial desert areas is particularly strategic, insofar as those processes faced a number of major difficulties, such as the limited geographic knowledge of this environment (due to its difficult access and consequent cost of exploration and occupation) and the imprecision with which the borders were defined in the boundary conventions, largely resulting from this knowledge gap. In addition to this, these processes had to contend with the spatiality of the indigenous populations, whose vision of space and boundaries reflected criteria of subsistence rather than political belonging. The second part will detail the role of cartography in the process of delimitation and demarcation of colonial Sahara from its partition in the late nineteenth and early twentieth centuries to its formal demarcation in the 1950s, which will be the main focus of this section. This study draws mainly on archival materials, most of them unpublished, consulted in Spain—at the *Archivo General de la Administración* (AGA) (General Administrative Archives), the *Archivo Cartográfico y de Estudios Geográficos del Centro Geográfico del Ejército* (ACEGCGE) (Cartographic and Geographical Studies Archive of the Spanish Army Geographic Centre), the *Instituto Geográfico Nacional* (IGN) (National Geographic Institute) and the *Biblioteca Nacional de España* (BNE) (National Library of Spain)—and in France—at the *Bibliothèque Nationale de France* (BNF) (National Library of France) and *Centre des Archives Diplomatiques de Nantes* (CADN) (Nantes Diplomatic Archives Centre).

5.2 Drawing Lines in the Sand: Cartography and Borders in Colonial Desert Territories

The place of cartography in African colonial partition and history has been the subject of many studies, some focusing on the maps produced in the processes of delimitation and demarcation of colonial territories (Braun 2024; Lefebvre 2015; Blais 2014). Recurring themes in this literature have included the notable difficulty of cartographic operations in some of these territories, the relationship of maps to the disputes arising in the process of conquest and territorial delimitation by the colonial empires (both between and within them), or indeed how maps contributed to the imposition—on indige-

[1] The main events, projects and debates surrounding the border delineation of the Spanish Sahara have been summarized by Martínez Milán (2012) and Raya-Collado (2023).

nous populations—of modern forms of territorial sovereignty. Within the framework of these concerns, borders in desert areas present a particular problem.

The analysis of deserts as geographic phenomena conducive to the separation of peoples, civilizations and political entities has been addressed by the classics in the literature on boundary-making (Curzon 1907; Holdich 1916; Fawcett 1918; Boggs 1940; Jones 1945). Just like seacoasts or major mountain ranges, large deserts were considered archetypes of 'natural frontiers'—that is, formidable zonal barriers considered extremely valuable for defensive purposes due to the significant difficulties they posed both for movement and, above all, for human settlement. However, the drawing and maintenance of linear borders in this type of environment pose a number of recurrent challenges, such as the harsh climatic conditions; the scarcity of inhabitants, stable settlements and natural topographical landforms that can serve as reference points; the dynamic, unstable and changing nature of sandy landscapes; and the difficulty of adapting this type of border to the traditional spatial patterns of native populations, in many cases characterized by nomadism. In addition, at the time when the claims and treaties establishing the colonial partition of these areas were being drawn up, there was little geographic knowledge of most of these spaces and, linked to this, they were perceived as being of low economic value. All of these reasons explain why colonial powers often chose to set geometric boundaries, whether astronomical (meridians and parallels) or mathematical (straight segments, circumference arcs, equidistant lines, etc.).

This link between geometric boundaries and deserts was already observed by seminal authors, such as the aforementioned Jones: 'Because of the sparse population, the paucity of naturally marked sites, and, in many cases, the rudimentary state of geographical information on deserts, geometrical boundaries are common in them'. (Jones 1945, p. 105). Yet these authors also warned of the many problems and challenges associated with such borders, which were typically described as 'artificial' in the literature of the time. Indeed, geometric boundaries proved useful (and above all convenient) for colonial governments that, from afar, were delimiting territories they knew little to nothing about and considered to be of little value. Yet they carried many risks and provoked conflicts when they had to be specified, demarcated and administered on the ground. 'It is tempting', wrote Jones, 'to place a ruler on a map and draw a line which is to be a boundary. Such a boundary, however, may be disharmonic with the facts of land and people. Widespread use of geometrical boundaries testifies to absentee, office-bound boundary-making' (Jones 1945, p. 151).

Jones himself outlined some of these problems, as had Holdich three decades earlier: the drawing of geometric boundaries required adequate geodetic knowledge, which was not always available at the time, and which could not be acquired or verified without fieldwork; the descriptions in the boundary treaties often lacked fundamental data to define the bearings of the lines (geodetic longitudes and latitudes are not the same as astronomical longitudes and latitudes, for example); the demarcation and maintenance of these lines could be very costly; and, above all, they could pass through natural or human phenomena whose division was highly problematic and inadvisable.

In the case of deserts, for example, a geometric boundary that crossed certain water-related landforms (such as oases, wells, mountainous areas)—which in these environments represent centres of human concentration—could lead to serious conflicts between the colonizing countries or between the colonizers and the local populations, just as would be the case when dividing up other highly valuable resources, like salt pans, mineral deposits or oil and natural gas fields. Similarly, linear borders drawn in this way could separate, for instance, a village from its tribal community of belonging, or cut off the migratory routes of nomadic groups, thus provoking their resistance and complicating the effective implementation and enforcement of the colonial border: as we quoted before, 'nothing but the necessity imposed by ignorance', Holdich went so far as to say, 'can justify the adoption of the straight line' (Holdich 1916, p. 170).

The issue of geometric boundaries in African deserts or in other desert areas that were divided by the colonial empires directly implicates cartography for its role in the appropriation and colonization of these spaces and, more specifically, the role of maps in boundary disputes. According to Gerald Blake, this relationship, broadly considered, can manifest in four main ways: 'First, they [maps] may be contributory factors, or even causes of the dispute; second, they may be used by the parties to promote their own position in a dispute; third, they are frequently among the most useful tools in dispute resolution; and, fourth, they may be used to illustrate the judgement of mediation or arbitration' (Blake 1995, p. 48).

All of these dimensions or manifestations are visible and recognizable in the processes of African border delimitation, as the case analyzed here will attempt to illustrate. It should also be recalled that African borders have traditionally been criticized, even stigmatized, as random 'a priori borders, drawn in Europe on indistinct maps with many blank spots and unknown territories, usually prior to any field reconnaissance'. (Foucher 2020, p. 289). As Donaldson explained, 'the initial partition of the African continent through the creation of inter-imperial boundaries was first expressed only in the geographical imagination of mapping [...]. African territory that was partitioned by the legally defined boundaries concluded between competing European powers during the so-called 'scramble' in the last decade of the nineteenth century was, in most cases, only conceived through the geopolitical lens of small-scale mapping' (Donaldson 2008, pp. 477–478). And as Basset summarized for the case of West Africa, 'maps were used as both instruments and representations of imperial power [...]. Like other forms of colonial discourse, such as speeches, books and sermons, maps promoted the appropriation of African space under the rhetoric of commerce and civilization. Maps facilitated conquest by their accuracy and selective content' (Basset 1994, p. 333).

In any case, in general, the treaties that, subsequent to the Berlin Conference, established the border delimitations of colonial Africa—and that, in some cases, were limited to defining the respective areas of influence of each country[2]—had to correct the rigid geometric lines to adapt them to the agreements previously reached with local authorities or to other needs expressed by the colonial representatives and by the commissions sent to survey and map these borders on the ground. In addition, some boundary treaties were rectified and adjusted locally after the fact, as—partly through the work of new boundary commissions—more accurate knowledge of the territory was acquired and the problems caused by the original delineations came to light. All of these adjustments, prior or subsequent to the treaties between the colonizing countries, explain why geometric lines were disrupted, diverted and distorted to incorporate settlements, rivers, roads and routes, grazing land, wells and water sources, and so forth into the same colony or partition, or to fix the border at a river course (Griffiths 1986).

As a matter of fact, in recent years, a number of studies have firmly challenged the arbitrariness and artificiality that conventional historiography attributes to Africa's colonial borders (Foucher 2014, 2020; Paine et al. 2025). Paine et al. (2025, p. 12) have calculated, for example, that 62 per cent of the 107 bilateral borders existing in Africa in 1960, before the major processes of decolonization began, were drawn along 'historical political frontiers' or natural geographic landforms (especially rivers or watersheds) many of which were also historical

[2] In this sense, it is important to remember that, with regard to colonial partition, the Berlin Conference limited its scope to establishing very general rules for claiming sovereignty or protectorate over coastal African territories, rather than defining concrete border lines. These rules were based on the principle of effective occupation of the territory, which was vaguely outlined in Articles 34 and 35 of the General Act of the Conference, signed on 26 February 1885.

political boundaries.[3] Likewise, although it is true that geometric boundaries (astronomical or otherwise) carried considerable weight in shaping the political and administrative map of colonial Africa (and, therefore, of postcolonial Africa, fundamentally built upon the former),[4] these authors have highlighted how, for the most part, such borders were established in arid and desert environments where, in general, population density was very low, there were few physical landforms of reference for drawing a border, and the value of the territory—and with it, its strategic interest—was considered minor by the states involved in the colonial process.[5]

Finally, with regard to the border delimitation of desert areas both in Africa and the Arabian Peninsula, several studies have revealed the problematic coexistence between the border model imported by the European colonizers and the spatial patterns of the predominantly nomadic native populations, many of whom were affected by the imposition of this model (Mizrahi 2009; Ouallet 2021). The colonial occupation of these spaces faced resistance from some of these populations—at times appeased by pacts reached with them, but also, at other times, crushed by the violence and military might of the colonial troops (Bonte 2012). The thesis that 'the colonial territorial logic and the idea of the border were opposed to the dynamics of the pastoral economy and the imperatives of nomadic pastoral routes' has been accepted as generally valid, although, when examined case by case, it has also been nuanced and corrected (Bonte 2012, p. 44).

This is partly because, until the European powers had demarcated and effectively controlled the border territories—which in some cases took several decades depending on the interests in play—the colonial borders functioned as relatively open and permeable zones in which the boundary line existed more on the map and on paper than on the ground, without significantly altering the mobility patterns of the local populations (López Bargados 2012). This is also because some studies have shown that certain colonial border or administrative delineations in the Sahara and the Sahel—as well as in other arid regions, within or outside Africa—took close account of the spatial distribution of the different local nomadic groups, which functioned as de facto precolonial political boundaries (Gagnol 2022).[6]

The fact that the geometric boundaries were closely related to the arid and semiarid environments does not mean that these boundaries ignored the anthropogeographic characteristics of the territory or were detached from regional and local political and socioeconomic dynamics. Far from it, the definition and materialization of these borders entailed conflicts and negotiations between various sociospatial actors, both exogenous and endogenous, with their differing interests and territorial visions. In this sense, beyond the agreements and treaties that legalized the African colonial partition on paper, the delimitation of the colonial desert spaces on the

[3] According to these authors, the historical political frontiers that were taken into account in the drawing of colonial borders include the borders of precolonial African states; those 'created by white settlements and states, in particular in southern Africa; non-European states such as Ottoman territories in the north and the resettlement colony of Liberia; and several decentralized groups, often nomadic groups whose grazing areas were used to determine intraimperial borders in areas of low population density' (Paine et al. 2025, p. 12).

[4] According to Paine, Qiu and Ricart-Huguet, geometric criteria were the primary criteria in 37 per cent of the 107 bilateral borders in force in Africa by 1960 (Paine et al. 2025, p. 15); and according to Foucher, geometric boundaries comprise 42 per cent of the total length of current African land borders (Foucher 2020, p. 289).

[5] According to the calculations by Paine, Qiu and Ricart-Hughet, '[a]mong the 27 [bilateral] borders located largely within deserts [by 1960], straight lines are the primary feature of 78%, compared to 24% for the 80 non-desert borders' (Paine et al. 2025, p. 15).

[6] Thus, for example, Gagnol has identified how the borders between Algeria and Niger, Algeria and Mali and Mali and Niger rely on the spatial distribution of the different Tuareg confederations present in the area, based in turn on the location and use of the principal water wells (Gagnol 2022). In desert regions of Asia, consultation with local populations in delineating boundaries was also frequent, as Boggs noted in 1940: 'if a boundary cramps the nomad and denies him a water access to the precarious resources of wells, water holes and grazing grounds, he is apt to disregard it until distant authorities reluctantly modify the frontier' (Boggs 1940, p. 137).

ground was a complex and protracted process that involved reducing what had previously been frontier *areas* into clear, precise and mappable boundary *lines*. While these borders may appear to be clearly delineated on maps, in practice they initially functioned as zones of transit, contact and exchange between the colonizing countries themselves and between the colonizers and the local populations—zones divided by lines invisible to all on the ground until these countries achieved real and effective dominion and administrative control over the territory. In this respect, particular emphasis should be placed on the role of cartography in the geometrization and administrative purification of desert frontiers, whose abstraction and scientific representation of space contributed to the visualization, materialization and institutionalization of a linear border that would not otherwise have developed in the desert.

5.3 From (Zonal) Frontiers to (Linear) Boundaries: Mapping and Border Issues in 'Spanish Sahara' (1884–1957)

5.3.1 The Colonial Partition of Western Sahara and Early Cartographic Disputes (1884–1950)

Spain's interest in the Sahara began at the end of the nineteenth century, on the occasion of the Berlin Conference. Eager to control the Saharan coast in order to maintain its sovereignty over the Canary Islands, on 26 December 1884, Spain notified the other states at the Conference that it had decided to 'take under its protection the territories of the west coast of Africa between the aforementioned West Bay and Cape Bojador and in which are located, in addition to the aforementioned sites, Las Puntas and La Bombarda [...]'. (AGA, doc. 1).[7]

[7] The length of the coastline covered by this territory totalled about 550 kms.

While Spain's presence in the territory was then limited to three sites on the coast—Villa Cisneros, Angra de Cintra and Cape Blanco—in April 1885 the French government, seeking to connect its possessions in the north of Africa with those in the west, asserted that it held rights to the territories between Cape Blanco and Cape Bojador. It was then that the French-Spanish negotiations for the colonial partition of Western Sahara began (Brownlie 1979, pp. 147–158).

The initial talks, which revolved around the delimitation of the Cape Blanco peninsula (Vilar Ramírez 1977, p. 84), in the south of the territory, and the border near the Iyil salt pans (Rodríguez Esteban 2008), in the east, resulted in the signing of the Boundary Convention of 27 June 1900, whose first article defined the southern and eastern borders of the Sahara using the geometric criterion characteristic of the delimitation of desert areas:

> On the coasts of the Sahara the boundary between the Spanish and French possessions will follow a line that, starting from the [...] point located on the west coast of the Cape Blanco peninsula [...], will proceed through the centre of the said peninsula, and then will rise northwards to meet the parallel of latitude 21°20′ north. The border will continue east along the parallel of latitude 21°20′ north until the intersection of this parallel with the meridian 15°20′ west of Paris (13° west of Greenwich). From this point, the demarcation line will follow in the N.W. direction extending between the meridians 15°20′ and 16°20′ west of Paris (13° and 14° west of Greenwich), a curve drawn so as to leave the region of the Iyil salt pans and its dependencies to France, maintaining a distance of at least 20 km from the outer limit of the said salt pans. From the meeting point of this curve with the meridian 15° west of Paris (13° west of Greenwich) the border will proceed as straight as possible to the intersection of the Tropic of Cancer and the meridian 14°20′ west of Paris (12° west of Greenwich) and will extend along the direction of the latter meridian in a northerly direction. (Cordero Torres, 1962, pp. 242–244)

Annexed to the text was a map[8] signed by the representatives of Spain and France, which

[8] The map, *Anexo n°2. Carta de delimitación de las posesiones españolas y francesas entre Cabo Blanco y Cabo Bojador* (Appendix no. 2. Map delimiting Spanish and French possessions between Cape Blanco and Cape Bojador), can be found in De Reparaz (1920, Appendix).

showed the shape and trajectory of the borders legally formalized in the treaty. This map, like the first article of the Convention, already reveals one of the main characteristics of colonial border delimitation in the desert: borders were imprecisely defined due to the scant geographic knowledge that the colonial powers had of this environment. The geometric outline of the boundary line, the predominance of empty spaces, the small scale of the map and the geographic imprecision of some expressions used in the agreement to describe the aforementioned line, such as 'in the N.W. direction', 'distance of at least' and 'as straight as possible', demonstrate the limited foothold of Spain and France in the Saharan hinterland when they defined its borders.

This explains why the northern border of the Sahara, whose delimitation was subject to the French-Spanish negotiations on the partition of areas of influence in Morocco, was not more precisely defined (Martínez Milán 2015). In the Boundary Convention of 3 October 1904, it was established that the line would start from the intersection of the meridian 12° west of Greenwich with the 26° parallel, from which it would continue eastward until meeting the 8°40′ meridian (Cordero Torres 1962, p. 76), while in the Boundary Convention of 27 November 1912 it was agreed that the border would extend from the intersection of the 8°40′ meridian with the 27°40′ parallel to the wadi of the Oued Draa, located in present-day Morocco (Cordero Torres 1962, p. 99) (Fig. 5.1). In other words, the system of parallels and meridians proved convenient and, above all, useful for delimiting a vast, remote territory which the colonial powers knew very little about.

This initial delimitation, however, did not translate into an immediate, effective demarcation on the ground. Instead, it took decades for the boundaries formalized in the Convention of 1900 to materialize.[9] And this is another of the main characteristics of colonial border demarcation in the desert: until the colonial powers actually occupied their colonies, the boundary line existed more on the map and on paper than on the ground.

In the case of Western Sahara, the borders agreed upon at the beginning of the century remained unknown to the colonial authorities and local populations until the 1950s. This was due to the late, limited and inconsistent occupation undertaken by France and, especially, by Spain. While France began to advance into the Mauritanian Adrar from 1909 onwards, when it managed to impose its authority over the tribes present in the region, Spain did not establish its first military post within the Sahara until the spring of 1934. Up to then, its colonial operations were based on the policy of 'peaceful penetration' developed by the first political-military governor of the Sahara, Captain Francisco Bens Argandoña, which involved gaining the favour of the territory's tribes through gift giving (Bens Argandoña 2016 [1947]). This policy, which had little to do with the territorial control inherent in a context of colonial occupation, thus proved insufficient for establishing a permanent presence in the Saharan hinterland. Fifty years after declaring its protectorate over the Sahara, the Spanish occupation was reduced to two coastal settlements—Villa Cisneros and La Güera, on the Cape Blanco peninsula—guarded by small detachments with no possibility of penetrating into the interior of the territory, and there was not even an official census of the colony's population.[10]

Consequently, the colonial powers had a very limited presence in their possessions, which explains why the first colonial maps contained numerous gaps and technical inaccuracies concerning the border delineation, on account of their scale and drafting methods. For example, the *Mapa del Sáhara español y*

[9] The borders of Western Sahara defined in the 1904 and 1912 Conventions deserve special treatment—beyond the scope of this chapter—as they are more linked to the geopolitics and history of French-Spanish colonization in Morocco.

[10] The first census, in 1950, yielded a figure of approximately 13,000 inhabitants, although this number proved quite inexact in later recounts, given the nomadic nature of much of the Sahrawi population (Estévez 2012). Already in 1954, Molina-Campuzano, the author of the first detailed study of population recounts in Spanish Sahara, estimated that there were approximately 30,000 nomadic Sahrawis in the colony (Molina Campuzano 1954).

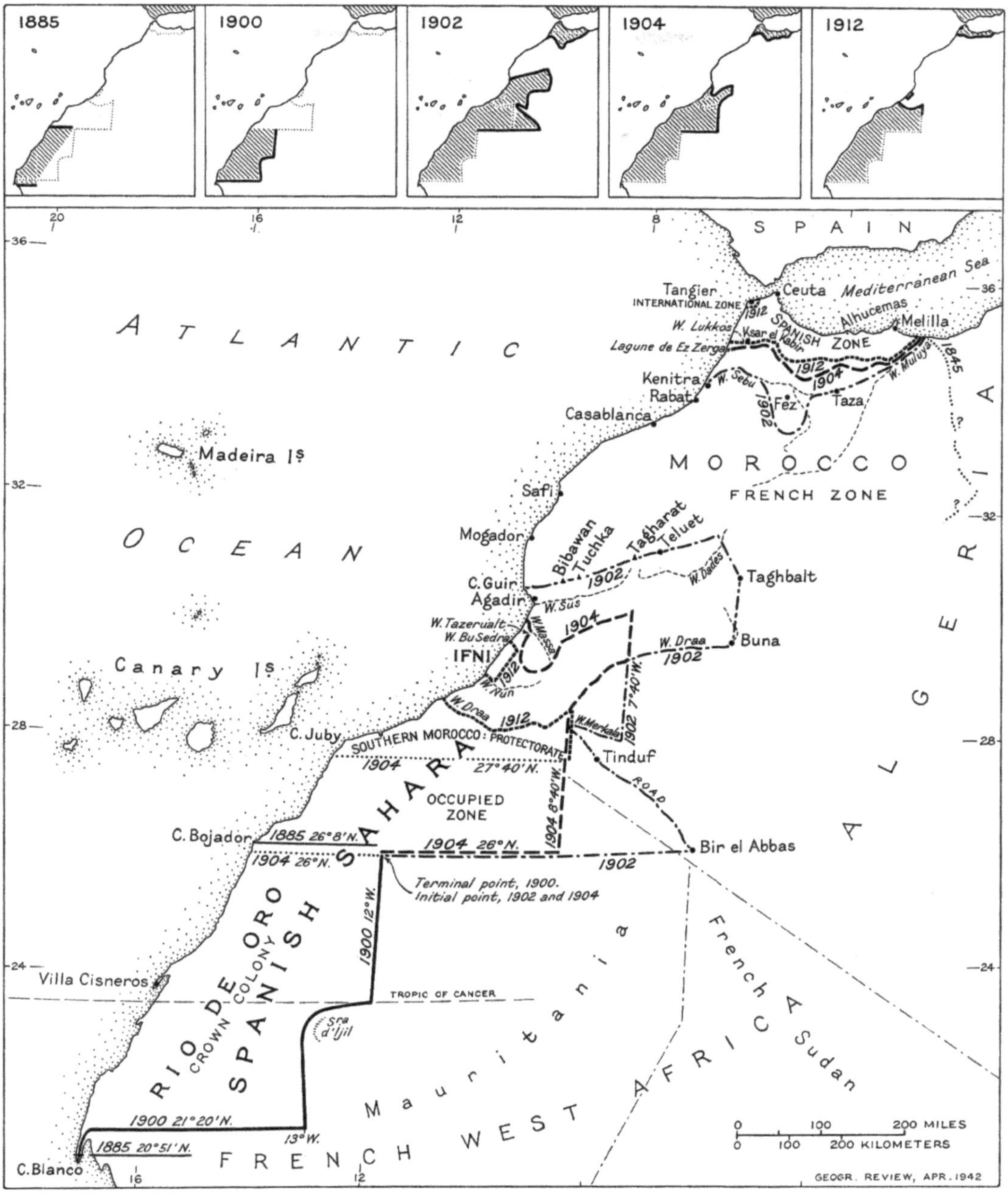

Fig. 5.1 Borders agreed for the Spanish Sahara upon in the Spanish-French Boundary Conventions of 1900, 1904 and 1912. The map also shows the territorial allocations that were considered (but ultimately rejected) between 1902 and 1904. By 1912, colonial Western Sahara consisted of three zones: *Río de Oro* (between parallels 21° 20′ N and 26° N), *Saguia el Hamra* (between parallels 26° N and 27° 40′ N) and *Cabo Juby* (between parallel 27° 40′ and the Oued Draa). *Source* Deasy (1942, p. 305)

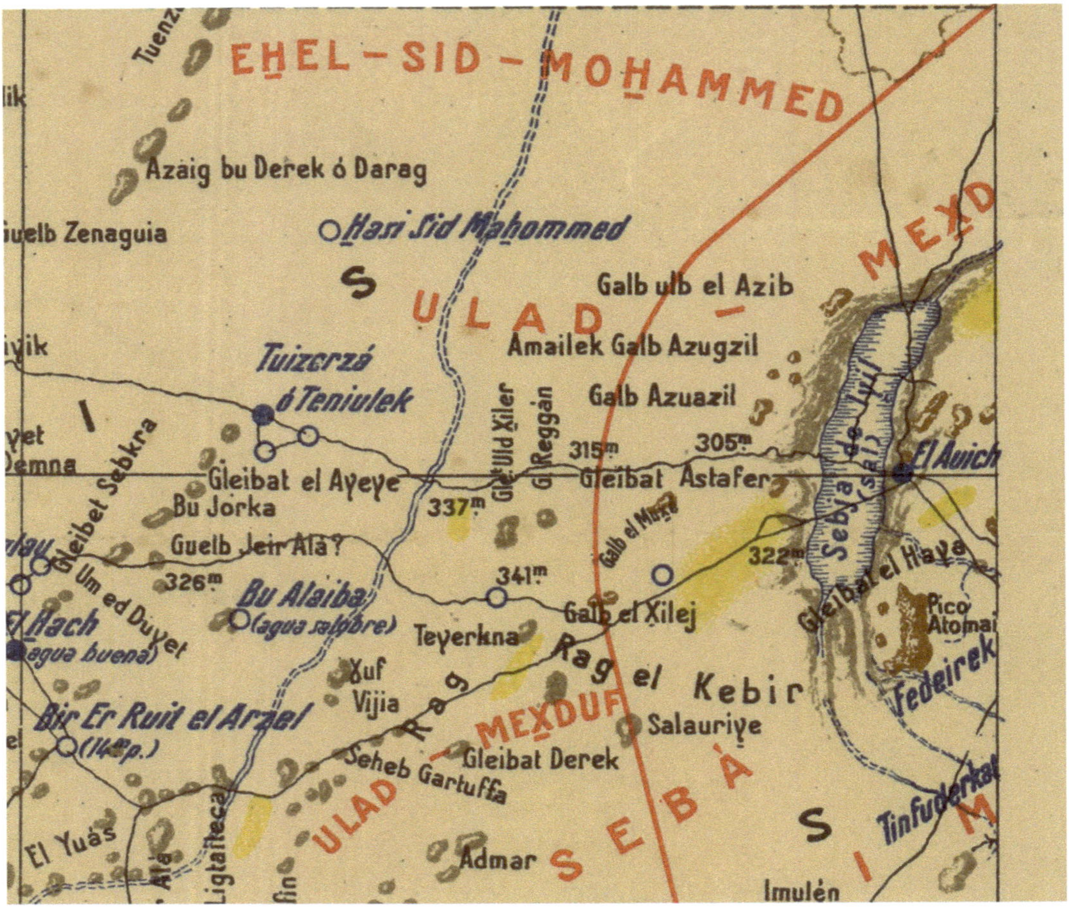

Fig. 5.2 *Mapa del Sáhara español y regiones inmediatas* (Map of Spanish Sahara and immediate regions) at 1:1,000,000 scale, by Enrique D'Almonte, published by the Royal Geographic Society of Madrid in 1914. Detail of the region surrounding the Iyil salt pans, located on the right-hand side of the map to the east of the boundary line coloured in red. These salt pans, attributed to France by the Boundary Convention of 1900, are erroneously shown to be crossed by the 23rd parallel. *Source* BNE, MR/33-41/1832

regiones inmediatas (Map of Spanish Sahara and immediate regions) on the millionth scale (1914), drafted by geographer and explorer Enrique D'Almonte and published in Madrid by the *Real Sociedad Geográfica* (Royal Geographic Society), together with an extensive geographic description of the territory (D'Almonte 1914), located the Iyil salt pans in the vicinity of the eastern border of the Sahara and crossed by the 23° parallel, when they actually lie below it (Fig. 5.2).

The following complete map of the Sahara available to Spain also failed to substantially improve knowledge of the territory and its borders, as it was essentially based on D'Almonte's map and not on a direct survey of the terrain.[11] Meanwhile, French cartography—the result of various expeditions by French troops to the Iyil salt pans since entering the Adrar—depicted these salt pans as connected to those of Um Dferat to the north of the 23°

[11] It was the *Plano del Sáhara español y regiones inmediatas* (Map of the Spanish Sahara and immediate regions) to the millionth scale, published by the General Government of the Sahara in 1934 and drafted from references, as indicated in the legend. Most of the information is taken from D'Almonte's map. Copies of it are available at the ACEGCGE.

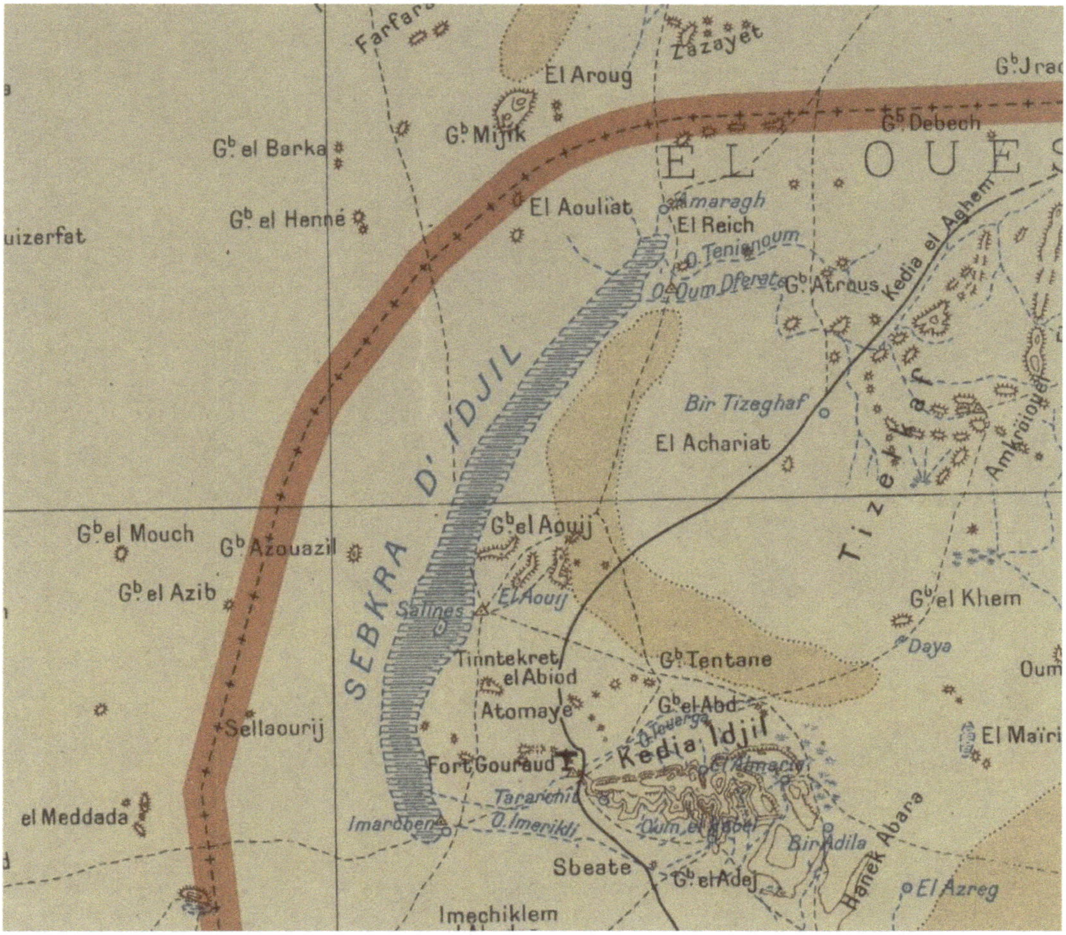

Fig. 5.3 Detail of the Iyil salt pans on sheet NF-28 (Port-Étienne) from the *Croquis de l'Afrique française* (Sketches of French Africa). The salt pans of Um Dferat (Oum Dferat in the map) are depicted to the north of the Iyil salt pans (Sebkra d'Idjil) and the 23° parallel. *Source* BNF

parallel (Figs. 5.3 and 5.4), when they are actually separate, and included in French territory places that were actually on the Spanish side of the border, as agreed in 1900: this was the case for the boundary wells at Agüenit, Mabruc and Miyec, whose almost permanent water supply made them strategic sites for controlling the movements of the tribes in the region.

It was not until the Spanish government occupied the interior of the Sahara starting in 1934 that its maps began to display a more accurate and precise representation of the territory and its borders (which soon clashed with those of the French maps). These advances led to the 1937 *Mapa del Sáhara español* (Map of Spanish Sahara) on the millionth scale, drafted using the data collected by the *Tropas Nómadas* (Nomad Troops; an auxiliary regiment to the colonial army in Spanish Sahara aimed at traversing the Saharan hinterland) (Mulero Clemente 1945)[12]; they also led to the *Mapa del África Occidental Española* (Map of Spanish West Africa) at 1:500,000 scale, drafted by the Geographic Service of the Spanish Army between 1943 and 1949, in fourteen sheets, whose compilation relied on extensive field research conducted with the collaboration of Sahrawi guides (Rodríguez Esteban 2011). Both maps were the first ones

[12] A copy of the 1937 map is available at the ACEGCGE.

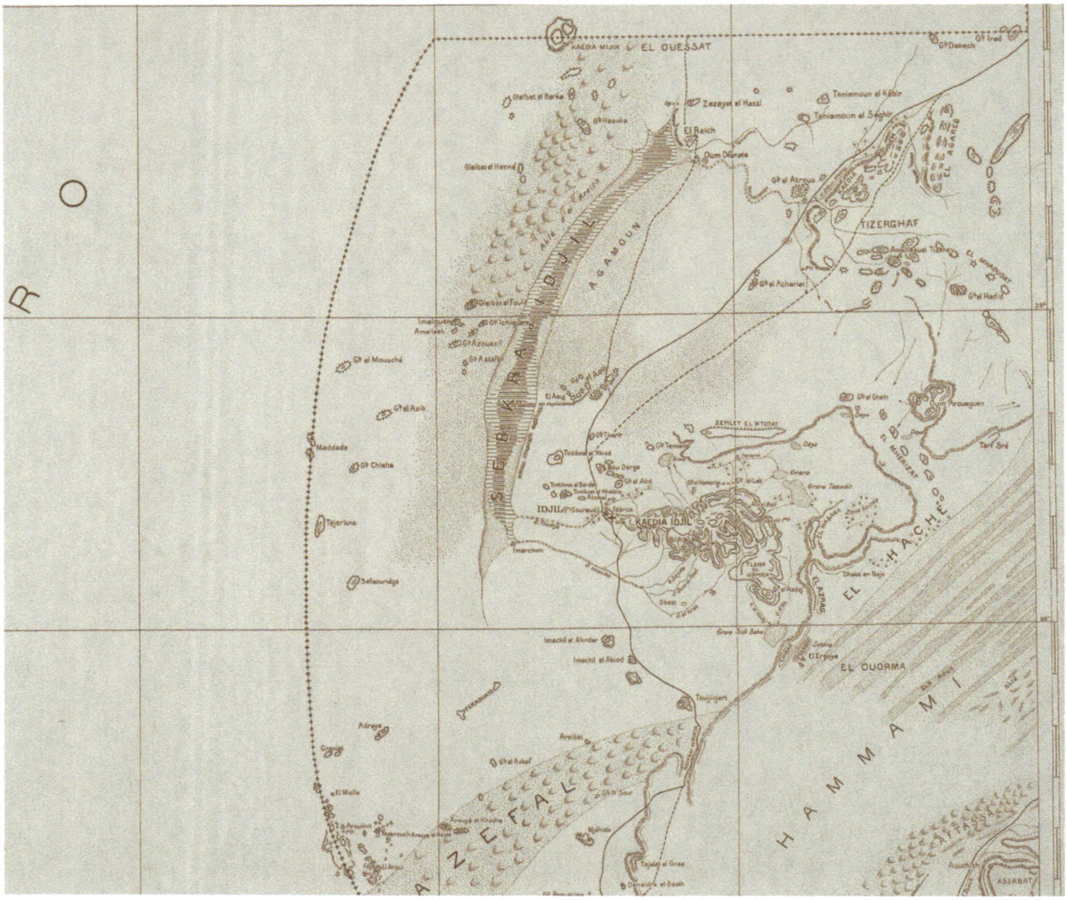

Fig. 5.4 Detail of the Iyil salt pans region on sheet NF-28 NE (Idjil, Fort Gouraud) from the *Croquis de reconnaissance des régions sahariennes* (Reconnaissance sketches of the Saharan regions). The boundary wells at Agüenit, Mabruc and Miyec are depicted as being on French territory. *Source* BNF

produced by the Spanish administration to depict the Iyil salt pans separately from the Um Dferat salt pans, and also to locate the boundary wells at Agüenit, Mabruc and Miyec in Spanish territory (Fig. 5.5).

As the colonial powers occupied the interior of their possessions, their geographic knowledge of them improved and, consequently, so did the accuracy of their maps, which became objects of dispute and instrumentalization to promote their own territorial visions on the drawing of borders that, although clear and linear in the maps and conventions, proved to be merely vague, invisible areas on the ground.

In this regard, the 1930s and 1940s maps portrayed as fixed, continuous, precise lines what on the ground were actually vast undefined areas—that is, zones of transit, contact and interaction between the colonial authorities, whose presence in the border regions was very limited due to the scant territorial establishment of their administrations and the harsh desert climate, which was ill-suited to the permanent establishment of the representatives of colonial operations. Thus, it could be argued that these maps, in the hands of officials arriving for the first time to the borders that their respective mother countries had established half a century earlier, served only to simulate colonial claims of territorial sovereignty, since on the ground the colonial powers had no permanent military presence and exercised no real dominion over

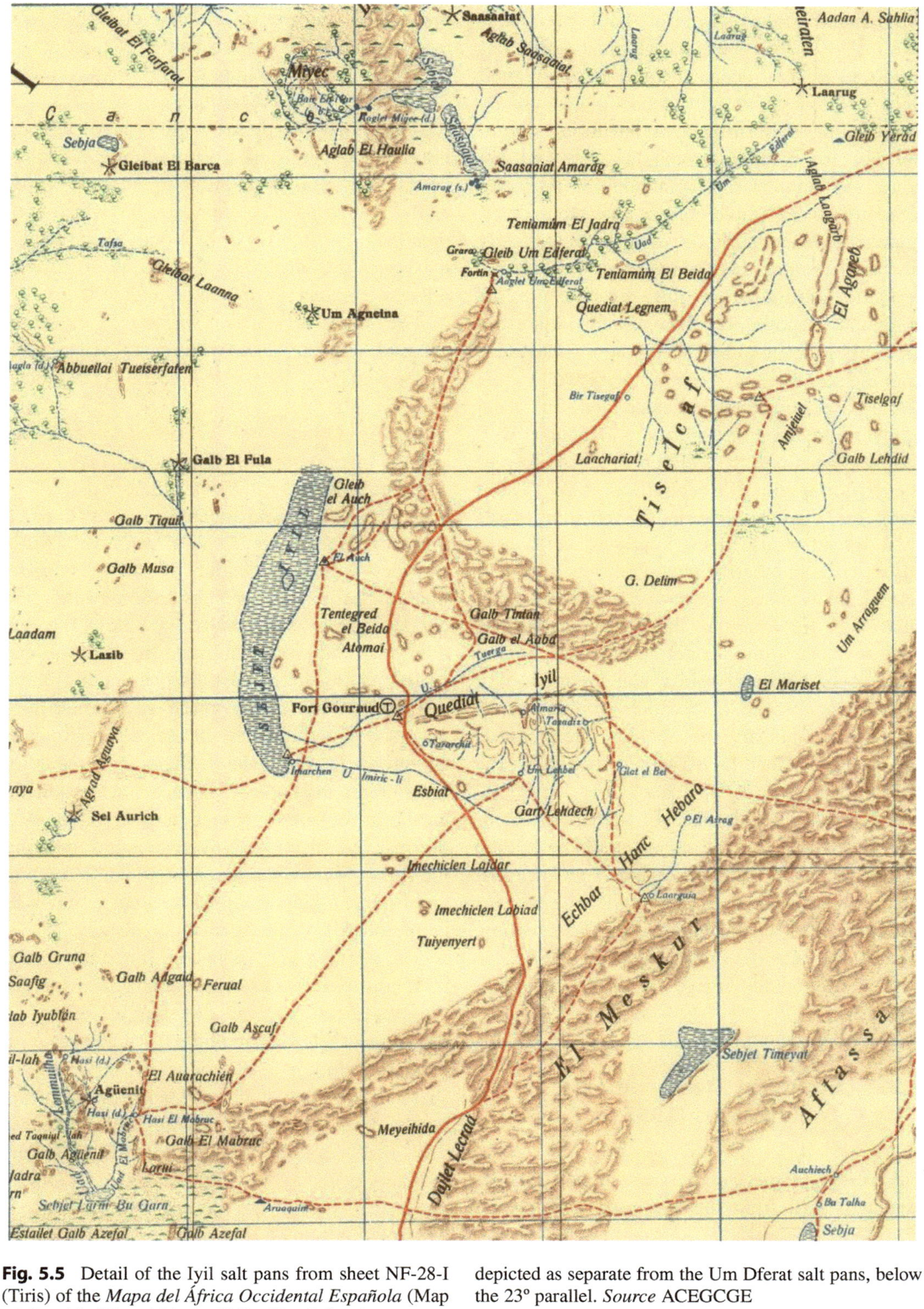

Fig. 5.5 Detail of the Iyil salt pans from sheet NF-28-I (Tiris) of the *Mapa del África Occidental Española* (Map of Spanish West Africa), 1951. The Iyil salt pans are depicted as separate from the Um Dferat salt pans, below the 23° parallel. *Source* ACEGCGE

the Saharan borders. Accordingly, the first maps that the Spanish and French colonial authorities produced as they occupied the interior of their Saharan possessions proved useful for visually representing their respective territorial claims, but not for resolving them, as they did not constitute detailed representations of the colonial borders that would make it possible to determine their exact delineation and hence to clearly divide sovereignty between the two powers.

Quite illustrative in this regard were the solutions proposed by the colonial authorities precisely to resolve the confrontations that arose over the sovereignty of the boundary wells while the borders remained undemarcated. Far from applying the logic of exclusivity associated with the modern, linear conception of the border as the territorial limit of a centralized power (a hallmark of colonial territoriality), in some cases they agreed to use them jointly (CADN, doc. 1),[13] while in others, France proposed to establish a kind of mixed or neutral zone around them. In Miyec, for example, the French authorities suggested 'as a provisional arrangement among neighbours on good terms, the creation of a mixed border zone [...] which would have as its eastern boundary the Spanish interpretation [of the border] and as its western boundary, the French one'. (ACEGCGE, doc. 1).

Ultimately, the absence of a modern scientific cartography of the region until the middle of the twentieth century prevented the colonial borders from being used as a privileged space for the exercise of exclusive territorial sovereignty, despite the fact that the maps asserted by these powers offered the image of clearly delimited territories.

5.3.2 Boundary Demarcation Commissions (1952–1957)

Nevertheless, this particular modus vivendi began to shift from the 1950s onwards, when the colonial powers discovered phosphate and iron ore along the eastern border of the Sahara, specifically in the vicinity of the so-called *curva de Iyil* (Iyil curve) (Martínez Milán 2021). The prospect that this desert region might contain important mineral resources prompted a change in the territorial logic hitherto adopted, especially by the Spanish administration, which until then had allocated fewer resources to the reconnaissance and exploration of its West African possessions than its French counterpart.

The *Mapa del África Occidental Española* (Map of Spanish West Africa) published in 1952 by the Spanish Directorate-General for Morocco and the Colonies (former body of the Prime Minister's Office—*Presidencia del Gobierno*) and the Institute of African Studies (*Instituto de Estudios Africanos*; IDEA), which was part of the Spanish National Research Council (*Consejo Superior de Investigaciones Científicas*; CSIC), provides a clear indication of the shift in the Spanish administration's attitude towards the demarcation of the Sahara (Fig. 5.6). The fact that this map was published by two official bodies just as the earliest discoveries of mineral resources were taking place in the territory underscores the urgency that seized the Spanish authorities to clarify the delineation of the colonial borders. It is no coincidence that the borders appear in red, a bold colour standing out against the rest, nor that the only two points marking the Iyil curve are the disputed wells at Agüenit and Miyec, which are, unsurprisingly, shown as being located within Spanish territory.

The defining of borders constituted the spatial prerequisite for the exploitation of resources discovered, and it was therefore necessary to determine their demarcation on the ground in order to eliminate situations of uncertain or shared sovereignty that could jeopardize projects for the development and exploitation of the territory. This was how Luis Carrero Blanco, undersecretary of the Spanish Prime Minister's Office, put it in February 1951: 'Given the indeterminacy of this border [the one projected near the Iyil salt pans] and considering that important mineralogical deposits have been found in its vicinity, which, by geological analogy of those

[13] This is the case for the Tifariti well, located in the vicinity of the parallel 26°, which serves as the north-eastern border.

Fig. 5.6 *Mapa del África Occidental Española* (Map of Spanish West Africa) at approx. 1:1,200,000 scale, published in 1952 by the Directorate-General for Morocco and the Colonies and the Institute of African Studies. *Source* IGN

lands over a wide expanse, may even straddle the border line, it is of the utmost importance to achieve its exact delimitation without delay'. (ACEGCGE, doc. 2).

The borders, consequently, had to materialize on the ground or, put another way, had to become *linearized*—that is, abandon their zonal character to be reduced to clear, fixed, precise lines in the landscape that would unequivocally divide the sovereignty of each power. Only in this way would they facilitate the territorial dominion necessary for occupying and exploiting the colonial possessions. And it was for this purpose that, at Spain's request, the

International Joint Commission for the Delimitation of Spanish Sahara with French Mauritania (*Comisión Mixta Internacional para la delimitación del Sahara español con la Mauritania Francesa*) was formed in February 1952.

The Commission, which according to Article 8 of the Boundary Convention of 1900 ought to have met within four months of the treaty's ratification, was not only a symbol of the change underway in the territorial logic of the colonial powers by the middle of the twentieth century; it was also the institution that enabled this transformation, due fundamentally to the importance it placed on scientific cartography in the process of border delimitation and demarcation. Within the Commission, maps not only served to show and legitimize the territorial claims of each colonial power, but also became an instrument for resolving them, as evidenced by the dispute that arose between the commissioners over the delimitation of the boundary line drawn opposite the Iyil salt pans.

Unlike the rest of the Sahara borders, the curve projected near Iyil did not follow the geometric criterion of parallels and meridians but, in accordance with the Convention of 1900, was to extend 'in a northwesterly direction' from its intersection with the 21° 20′ parallel (the southern border) following a curve between the meridians 13° and 14° west of Greenwich, which left to France the region of the Iyil salt pans and its dependencies. This curve was to be maintained 'at least' twenty kilometres from the outer boundary of the salt pans and, from its intersection with the meridian 13° west of Greenwich (the eastern border), it was to proceed 'as straight as possible' to the intersection of the Tropic of Cancer and the meridian 12° west of Greenwich. The imprecision of the expressions used to define the boundary line—'in a northwesterly direction', 'at least', 'as straight as possible'—as well as the region's strategic value due to its wells and mineral resources, underpinned the French proposal that 'the question of delimitation of the Saharan border should not be dealt with according to a strict, European-style standard, but particular local agreements should be reached to facilitate the life of the residents of the territory'. (ACEGCGE, doc. 3).

According to the French delegation, the Convention of 1900 'was to be considered a theoretical instrument whose rigidity should be relaxed' (ACEGCGE, doc. 4) because 'the lines that have been followed are purely theoretical and have no relation to the terrain. And since these lines could not be traced by boundary markers, because these would be ephemeral and difficult to locate, a broken line following natural boundaries would be appropriate.' (ACEGCGE, doc. 5). The French commissioners therefore proposed that the border near Iyil should pass 'through the Auadi well, along the peaks west of Agüenit, leaving the wells of Agüenit and Mabruc to France, [as well as through] Galg Gruna, El Meddada, Laadah, Tuiserfaten, Laamna, the Miyec well, Saasaiat (north of the Tropic of Cancer), Laaruc and [the] crossing of the Tropic with the 12° meridian' (ACEGCGE, doc. 4).

On the surface, the French proposal seemed to reflect a greater sensitivity to local contingencies in the Iyil region than what had been anticipated in the Convention of 1900, whose negotiators were unaware of what the French authorities would discover when they penetrated deeper into the Sahara: that it was a traditional nomadic area for the indigenous tribes due to the presence of major wells. However, the French proposal on the map revealed a different reality: France's proposed border delineation diverged considerably from the one legally formalized in 1900, to the detriment of Spain, which saw its territory reduced in a strategic region (Fig. 5.7).

The main differences between the French proposal and what had been agreed upon in 1900 revolved precisely around the vaguer terms of the treaty. The French commissioners based their proposed delineation on the interpretation of the expression 'northwesterly direction' as being an absolute northwest. They also argued that the reference to the boundary line proceeding 'at least' twenty kilometres from the outer boundary of the salt pans did not imply that the distance had to be exact, but could be approximate. In addition, French commissioners contended

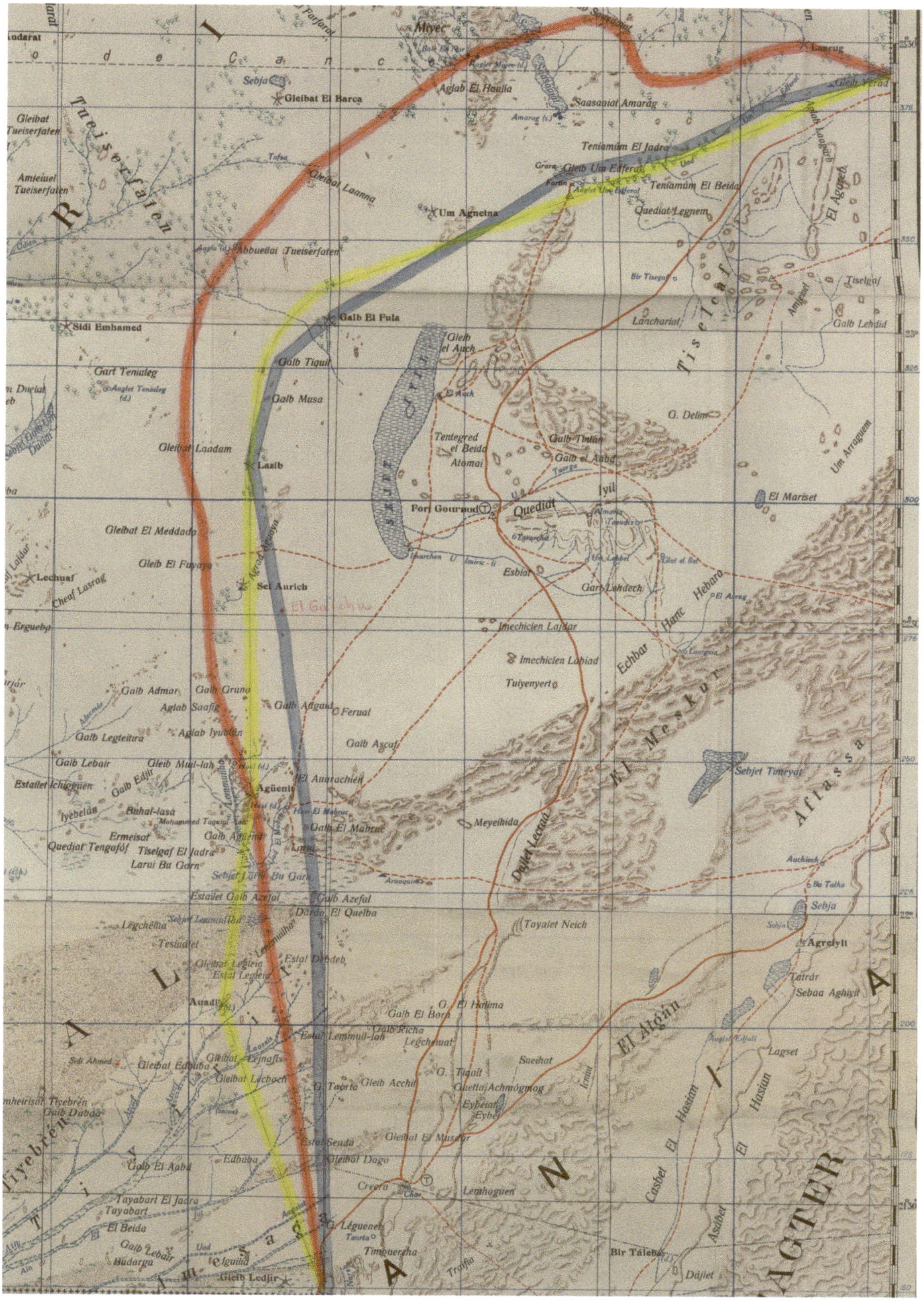

Fig. 5.7 Proposed French (red) and Spanish (blue) border compared to the line agreed upon in the Convention of 1900 (yellow) on sheet NF-28-I (Tiris) of the *Mapa del África Occidental Española* (Map of Spanish West Africa) at 1:500,000 scale, published by the Geographic Service of the Spanish Army in 1951 (1st edition). *Source* ACEGCGE

that the Iyil salt pans were connected to the Um Dferat salt pans, so that the expression 'as straight as possible' should not be translated to mean a straight line, but rather that 'this curve should be smoothed out'—meaning that the curve should cover the whole region of the salt pans, not just the area opposite them (ACEGCGE, doc. 5).

Most striking about the French delegation's proposal, however, was not that it proposed an openly different border delineation than the one legally agreed upon in 1900, but that it did so in the absence of a detailed, scientific map of the Iyil border region that would allow it to legitimize its interpretation of the treaty. Furthermore, the French delegation acknowledged to the Spanish commissioners that it had 'cartographic documents of small scale and mediocre value [...] to perform the comparison of documents as is the general rule in these cases' and that 'the documents at the disposal of the Spanish Delegation are superior.' (ACEGCGE, doc. 5). It therefore accepted that the boundary line near Iyil should be debated on the basis of the Spanish cartography, namely sheet NF-28-I corresponding to the Tiris region on the 'Map of Spanish West Africa' at 1:500,000 scale, discussed above (Fig. 5.5).

Indeed, the cartography available to the Spanish commissioners was superior because it was the first modern scientific map of the Western Sahara, produced by the Geographic Service of the Spanish Army in 1943–1949. The increased scale and improved accuracy and detail of the 'Map of Spanish West Africa' enabled the Spanish delegation to justify and legitimize its own interpretation of the boundary line near Iyil, which was much closer to what had been agreed upon in 1900. The Spanish delegation stated:

> The Treaty is described as being rigid and the lines it draws as boundaries suffer from being theoretical, cutting off natural regions. We cannot accept this, since the French delegation has accepted [...] the meridians and parallels that appear in the Treaty. These cut through all the natural regions, all the topographical landforms existing in the terrain and, consequently, the straight or curved lines that the Treaty imposes cannot be accused of being either rigid or theoretical. All must be respected with regard to the Iyil salt pans (ACEGCGE, doc. 5).

The Spanish commissioners thus rejected the geographic criterion claimed by the French delegation and proposed that the boundary line near Iyil conform as closely as possible to the geometric criterion stipulated in the Convention of 1900 for the rest of the boundary. But, unlike the French delegation, the Spanish delegation used its own cartography to defend its position. Relying on the data recorded on the territory in the Spanish West Africa map and on the geodetic and topographical work conducted by the Spanish technicians between March and May 1952 (ACEGCGE, doc. 6), the delegation presented two graphics that dispelled any doubts about the course the border should follow near Iyil (Figs. 5.8 and 5.9).

Whereas the first graphic showed that, if the border followed an exact northwesterly direction, it could be up to six times farther from the Iyil salt pans—120 kms—than anticipated in the 1900 agreement—at 20 kms distance—the second graphic demonstrated that the Iyil salt pans were less extensive than the French maps suggested, because they were not connected to the Um Dferat salt pans, so the boundary line should curve only as far as its intersection with the 13° meridian, from which it would proceed in a straight line to the crossing of the Tropic of Cancer with the 12° meridian.

Confronted with this graphic evidence, the French delegation ultimately accepted the Spanish interpretation, such that the Joint Border Commission agreed that, starting from the intersection of the 21° 20′ parallel with the 13° meridian:

> The border remains defined by *straight lines* [emphasis in original] that successively join the highest points of the following elevations: Galb Azefal, El Gaicha, Lazib, Galb Musa, Galb Tiquit, Galb El Fula, Gleib Um Edferat, Teniamun El Jadra, Gleib Lemhar, and Gleib Yerad, from which it proceeds *in a straight line* to the intersection of the Tropic of Cancer with the meridian 12° west of Greenwich, which then constitutes the boundary to the north (ACEGCGE, doc. 7).

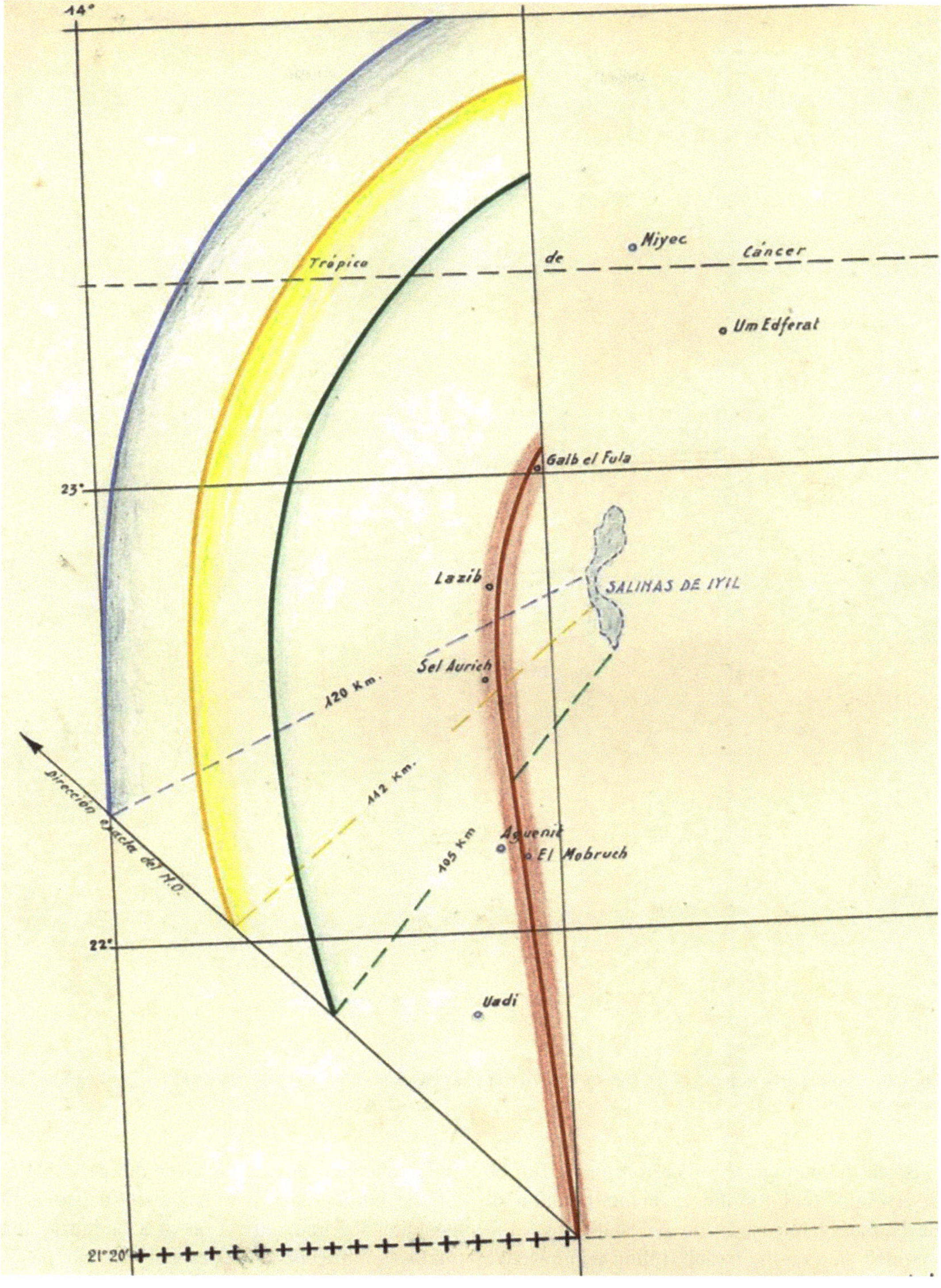

Fig. 5.8 Comparative graphic of the course the border would take near Iyil depending on the interpretation of the expression 'northwesterly direction'. *Source* ACEGCGE

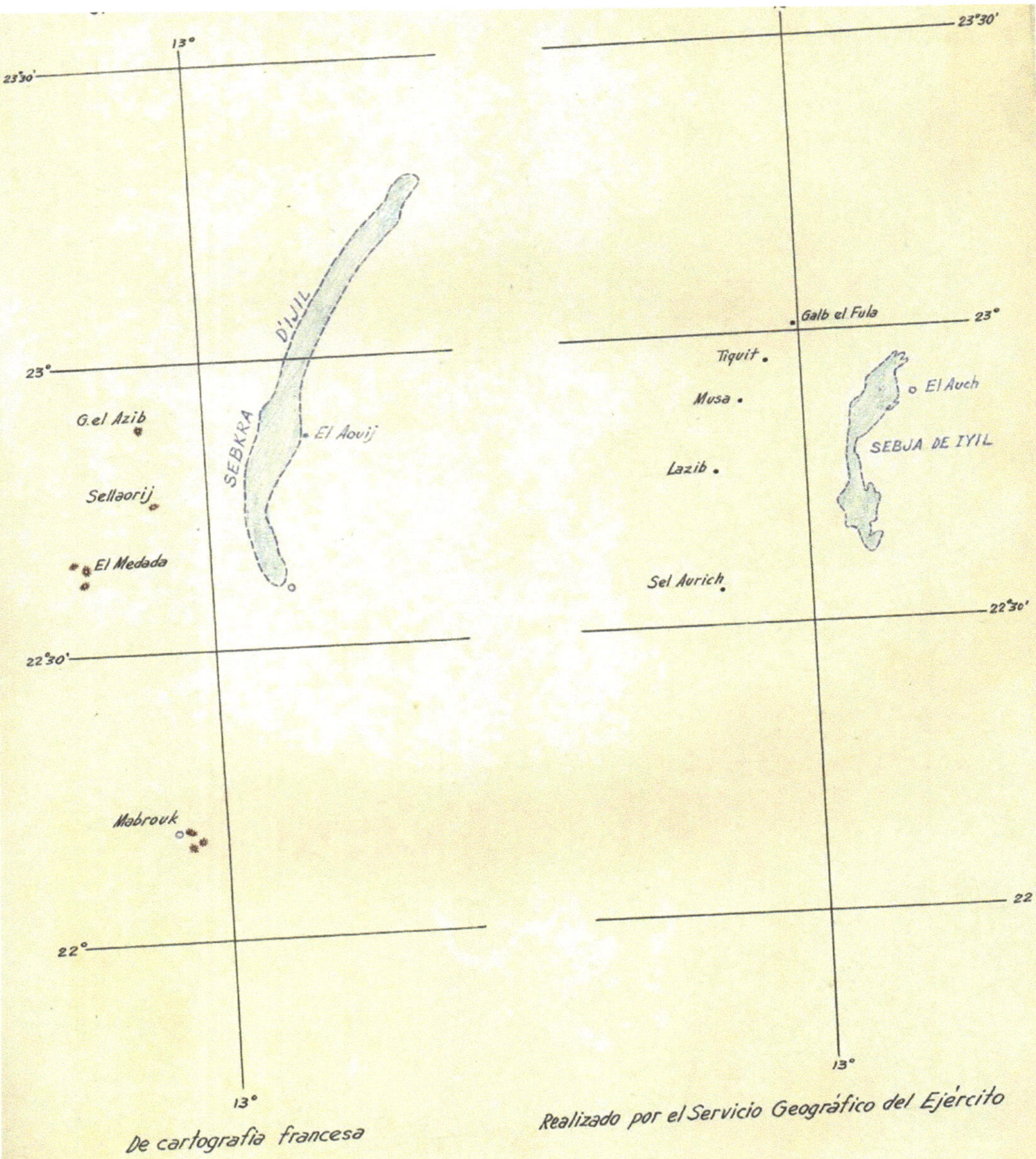

Fig. 5.9 Comparative graphic of the size of the Iyil salt pans based on French (left) and Spanish (right) cartography. *Source* ACEGCGE

Scientific cartography therefore proved to be an essential tool for the exact delimitation of the boundary line near the Iyil salt pans, as it provided data on the region, filling a knowledge gap on it and, consequently, resolving uncertainties about its demarcation. In this sense, the maps used by the Spanish commissioners not only reflected the dispute over the delimitation of the border near Iyil, but also shaped it by enabling the geometrization of a boundary line that was imprecise on paper and invisible on the ground. This process culminated in the demarcation of the borders between Spanish Sahara and French Mauritania in 1957, in which Spanish maps were also used as the primary reference, ending the conflicts among the two colonial powers over the sovereignty of the wells in the area by demonstrating that they were clearly located in Spanish territory.[14] Consequently, once the border demarcation was complete, the Spanish government decided not to authorize the French military to use the Agüenit and Mabruc wells for water supply (AGA, doc. 2). In this way, places that for years had represented neutral or mixed-sovereignty zones were now divided under the logic of exclusivity associated with the modern, linear conception of the border that was introduced and enabled by the maps of the Joint Boundary Commission.

5.4 Concluding Remarks

The delimitation of colonial Western Sahara, as discussed in this chapter, was a complex and protracted process that would not have been possible without the help of cartography. Because the desert environment of the territory hindered its exploration and occupation from the end of the nineteenth century, France and Spain agreed on geometric borders that proved to be imprecise on paper and invisible on the ground.

For most of the European occupation, the borders formalized in boundary conventions essentially constituted broad zones of transition and interaction between colonial troops themselves, as well as between these troops and the nomadic tribes present in the region, whose exact sovereignty was uncertain due to the lack of effective demarcation. This caused the territorial authorities to clash—with maps thrown into the fray—over control of strategic border sites in the desert, notably wells and to develop a modus vivendi much more marked by pragmatism than by the modern dictates of territorial sovereignty, linked to linear borders and centralized forms of governance.

It was not until the establishment of the demarcation commissions in the 1950s, and the use of detailed, scientific border maps, that these boundaries were inscribed on the ground as true lines of separation. Cartography, in this sense, proved to be a decisive tool for the development of modern territoriality in the Western Sahara region, as its scientific representation and abstraction of space contributed to the geometrization and administrative purification of colonial borders. Ultimately, the maps used by the French-Spanish commissions reduced to a clear, continuous, precise line what on the ground had always been broad areas of imprecise frontiers; maps thus put an end to the zonal character of desert borders and, with it, to situations of uncertain or shared sovereignty.

The case study examined here ultimately makes it possible to reflect on the difficulties and conflicts associated with the establishment of linear borders in desert environments and the role played by cartography in this process. But it also prompts us to consider, in future research, two other questions of significant interest for the history of cartography and the formation of modern states in these spaces. On the one hand, the relationship between the local populations and the territoriality of the colonial powers, whose linear borders often proved insufficient to prevent the movements of the region's nomadic tribes. On the other hand, the paradoxical assertion of colonial territoriality and linear borders not only by the states that have emerged in this region since decolonization (to the north of Western Sahara, Morocco, independent since 1956; and to the south, Mauritania, independent since 1960), and which, since 1975, have been contending for sovereignty over the former Spanish colony, but also of the Sahrawi people themselves when they began to demand their independence.

[14] For this purpose, in addition to using the sheets of the *Mapa del África Occidental Española* corresponding to the borders, the Geographic Service of the Spanish Army drafted, in 1957, a detailed map of the Iyil region, at 1:400,000 scale, *Mapa de la región fronteriza de la zona meridional del Sahara español con Mauritania. Región de la salina de Iyil* [Map of the border region of the southern zone of Spanish Sahara with Mauritania. Iyil salt pans region], available at the ACEGCGE.

Archival Sources

ACEGCGE (Archivo Cartográfico y de Estudios Geográficos del Centro Geográfico del Ejército, Madrid), doc. 1: Internal memo no. 28 on boundaries between Río de Oro and Mauritania [Nota de información interior nº 28 sobre límites entre Río de Oro y Mauritania], Sidi Ifni, 20 January 1949 (ACEGCGE, SAH. Shelf 6, T1, Box 1).

ACEGCGE, doc. 2: Memo from the Director General of Morocco and the Colonies to the Head of the Geographic Service of the Army on the delimitation of the Iyil salt pans [Escrito del Director General de Marruecos y Colonias al Jefe del Servicio Geográfico del Ejército sobre la delimitación de las salinas de Iyil], Madrid, 27 February 1951 (ACEGCGE, Ifni-Sahara (A), Box 391.

ACEGCGE, doc. 3: Proceedings of the International Joint Commission [Acta de la Comisión Mixta Internacional], San Sebastián, 28 July 1952 (ACEGCGE, SAH. Shelf 6, T2, Box 6).

ACEGCGE, doc. 4: Proceedings of the Restricted Technical Commission [Acta de la Comisión Técnica Restringida], San Sebastián, 29 July 1952 (ACEGCGE, SAH. Shelf 6, T2, Box 6).

ACEGCGE, doc. 5: Proceedings of the International Joint Commission with the presentation of the Spanish thesis by Lieutenant Colonel Aparicio [Acta de la Comisión Mixta Internacional con la exposición de la tesis española por el Teniente Coronel Aparicio], San Sebastián, 31 July 1952 (ACEGCGE, SAH. Shelf 6, T2, Box 6).

ACEGCGE, doc. 6: Memo from the Head of the Boundary Commission to the Director General of Morocco and the Colonies [Escrito del Jefe de la Comisión de Límites al Director General de Marruecos y Colonias], Madrid, 16 May 1952 (ACEGCGE, SAH. Shelf 6, T1, Box 1).

ACEGCGE, doc. 7: Text of the agreement on the delimitation and demarcation of the boundary of the southern zone of the Spanish Sahara with French Mauritania [Texto del acuerdo sobre delimitación y amojonamiento de la frontera de la zona meridional del Sahara español con la Mauritania francesa], San Sebastián, 1 August 1952 (ACEGCGE, SAH. Shelf 6, T2, Box 6).

AGA (Archivo General de la Administración, Madrid), doc. 1: Circular addressed to the Representatives of Spain abroad, informing them that H.M. the King has taken under his protection the territories of the west coast of Africa, between Cape Bojador and the West Bay [Circular dirigida a los Representantes de España en el extranjero, participándoles que S.M. el Rey ha tomado bajo su protección los territorios de la costa occidental de África, comprendida entre el Cabo Bojador y la Bahía del Oeste], Madrid, 9 January 1885 (AGA, IDD(15) 017, File no. 6, Box 81/00381).

AGA, doc. 2: Memo from the Director General of African Posts and Provinces on the use of border wells [Escrito del Director General de Plazas y Provincias Africanas sobre la utilización de pozos fronterizos], Madrid, 25 April 1957 (AGA, IDD(15) 23, Box 1).

BNE Biblioteca Nacional de España, Madrid.

CADN (Centre des Archives diplomatiques de Nantes, Nantes), doc. 1: Note verbale on Franco-Spanish relations at Tefariti [Note verbale sur les relations franco-espagnoles à Tefariti], Paris, 9 July 1945 (CADN, 183PO/1/82).

IGN Instituto Geográfico Nacional, Madrid.

References

Basset T (1994) Cartography and empire building in nineteenth-century West Africa. Geogr Rev 84(3):316–335. https://www.jstor.org/stable/215456

Bens Argandoña F (2016) [1947] Veintidós años en el desierto: mis memorias y tres expediciones al interior del Sáhara. Athenaica, Seville

Blais H (2014) Mirages de la carte. L'invention de l'Algérie coloniale. Fayard, Paris

Blake G (1995) The depiction of international boundaries in topographic maps. IBRU Bound Secur Bull 3(1):44–50. https://www.durham.ac.uk/media/durham-university/research-/research-centres/ibru-centre-for-borders-research/maps-and-databases/publications-database/boundary-amp-security-bulletins/bsb3-1_blake.pdf

Boggs W (1940) International boundaries. A study of boundary functions and problems. Columbia University Press, New York

Bonte P (2012) Frontières coloniales et parcours nomades. L'enjeu pastoral dans l'Ouest saharien. Hespéris-Tamuda XLVII:39–78

Braun LF (2024) Cartography in colonial Africa. In: Oxford research encyclopaedia of African history

Brownlie I (1979) African boundaries: a legal and diplomatic encyclopaedia. Hurst & Co., London

Cordero Torres JM (1962) Textos básicos de África, vol 2. Instituto de Estudios Políticos, Madrid

Curzon G (1907) Frontiers. Clarendon Press, Oxford

D'Almonte E (1914) Ensayo de una breve descripción del Sáhara español. Boletín de la Real Sociedad Geográfica LVI:129–348

Deasy GF (1942) Spanish territorial boundary changes in Northwest Africa. Geogr Rev 32(2):303–306. https://doi.org/10.2307/210278

De Reparaz G (1920) Aventuras de un geógrafo errante. Primera Parte: Soñando con España. Ferd. Wyss, Bern

Donaldson JW (2008) Pillars and perspective: demarcation of the Belgian Congo-Northern Rhodesia boundary. J Hist Geogr 34:471–493. https://doi.org/10.1016/j.jhg.2007.11.005

Estévez P (2012) Censos, identidad y colonialismo en el Sáhara español (1950–1974): la imaginación numérica de la nación española. Papeles del CEIC 89:1–34. ISSN 1695-6494

Fawcett CB (1918) Frontiers: a study in political geography. Clarendon Press, Oxford

Foucher M (2014) Frontières d'Afrique. Pour en finir avec un mythe. Débats-CNRS Éditions, Paris

Foucher M (2020) African borders: putting paid to a myth. J Borderlands Stud 35(2):287–306. https://doi.org/10.1080/08865655.2019.1671213

Gagnol L (2022) Géohistoire des frontières sahariennes. L'héritage nomade enseveli sous le murs de sable. Bulletin de l'Association de Géographes Français 99(1):53–75. https://doi.org/10.4000/bagf.9030

González Bueno A (2008) De la costa atlántica a la *sebja* de Iyil: la expedición de Francisco Quiroga, Julio Cervera y Felipe Rizzo por el Adrar en el verano de 1886. In: Rodríguez Esteban JA (ed) Conmemoración de la expedición científica de Cervera-Quiroga-Rizzo al Sáhara Occidental en 1886. CSIC, Madrid, pp 99–148

Griffiths I (1986) The scramble for Africa: inherited political boundaries. Geogr J 142(2):204–216. https://www.jstor.org/stable/634762

Holdich T (1916) Political frontiers and boundary making. MacMillan, London

Jones SB (1945) Boundary-making: a handbook for statesmen, treaty editors and boundary commissioners. Carnegie Endowment for International Peace, Washington, D.C.

Lefebvre C (2015) Frontières de sable, frontières de papier. Éditions de la Sorbonne, Paris

López Bargados A (2012) Perfiles del 'Far West' sahariano: el papel de la frontera colonial en el Sahel atlántico (Sáhara Occidental, Mauritania). Hespéris-Tamuda XLVII:19–38

Martín Escorza C (2008) La cartografía antigua del Sáhara Occidental. In: Rodríguez Esteban JA (ed) Conmemoración de la expedición científica de Cervera-Quiroga-Rizzo al Sáhara Occidental en 1886. CSIC, Madrid, pp 229–249

Martínez Milán J (2012) L'évolution de la conception frontalière du colonisateur franco espagnol et ses répercussions sur l'Ouest Saharien, 1900–1960. Hespéris-Tamuda XLVII:11–18

Martínez Milán J (2015) España y Francia en el sur de Marruecos: intereses, delimitación de fronteras y contencioso territorial, 1900–1912. Estudios de Asia y África 50(1):117–139. ISSN 2448-654X

Martínez Milán J (2021) Los recursos minerales y su explotación en el Sahara Occidental, 1942–1975. Ayer 122(2):243–271. ISSN 1134-2277

Mizrahi JD (2009) De la région frontière à la ligne frontière. Les confins méridionaux de la Syrie de la fin de l'Empire ottoman au début des Mandats. Vingtième siècle. Revue d'histoire 3(103):77–90. https://doi.org/10.3917/ving.103.0077

Molina Campuzano M (1954) Contribución al estudio del censo de población del Sahara español. CSIC, Madrid

Mulero Clemente M (1945) Los territorios españoles del Sáhara y sus grupos nómadas. El Siglo, Las Palmas de Gran Canaria

Ouallet A (2021) Frontières et projets territoriaux Exemple de l'espace sahélo-saharien. L'espace Géographique 50(1–2):101–118. https://doi.org/10.3917/eg.501.0101

Paine J, Qiu X, Ricart-Hughet JR (2025) Endogenous colonial borders: precolonial states and geography in the partition of Africa. Am Polit Sci Rev 119(1):1–20. https://doi.org/10.1017/S0003055424000054

Raya-Collado A (2023) Las fronteras coloniales del Sahara Occidental. Reflexiones sobre su trazado y materialización (1900–1960). In: García D, García-Álvarez J, Curnis M (eds) Fronteras del mundo hispánico. Ideas, conflictos, identidades. Sílex, Madrid, pp 165–180

Rodríguez Esteban JA (2008) Equívocos, mares interiores, salinas y cambios en las fronteras: consideraciones sobre la expedición y sus resultados. In: Rodríguez-Esteban JA (ed) Conmemoración de la expedición científica de Cervera-Quiroga-Rizzo al Sáhara Occidental en 1886. CSIC, Madrid, pp 251–268

Rodríguez Esteban JA (2010) La cartografía del Sáhara Occidental. In: Ortega N, García-Álvarez J, Mollá M (eds) Lenguajes y visiones del paisaje y del territorio. UAM, Madrid, pp 483–496

Rodríguez Esteban JA (2011) El mapa del África Occidental Española de 1949 a escala 1:500.000: orgullo militar, camelladas y juegos poéticos saharauis. Cybergeo: Eur J Geogr 516. https://doi.org/10.4000/cybergeo.23461

Vilar Ramírez JB (1977) El Sahara Español. Historia de una aventura colonial. Ediciones Sedmay, Madrid

Part II

Circulating and Communicating the Line

6 Milan and Paris in the Passo del Jacuy. The Circulation and Validation of Cartographic Images in a Debate Between Iberian Officials from Mid-eighteenth Century Río de la Plata

Lucía Rodríguez Arrillaga

Abstract

This chapter discusses the forms of transimperial circulation and validation of cartographic images mobilised by Iberian officials tasked with establishing their positions in the debates for the delimitation of the domains of their respective monarchs in the Río de la Plata in the mid-eighteenth century. From the analysis of the correspondence and reconstruction of the paths followed by the printed and handwritten maps referred to in these debates, it is possible to establish the criteria of authority and hierarchy that operated with respect to the reliability of their sources of knowledge. The chapter tries to show how the contenders mobilised this hierarchy in a biased way, with some privileging the D'Anville map printed in Paris in 1748 or what the 'geography curious' from Milan had to say about territories an ocean away over maps produced locally by 'natives raised without any science or doctrine' or maps produced by officials of the respective monarchies.

Keywords

Río de la Plata · Mid-eighteenth century · Demarcation debates · Circulation of images · Validation of images · Hierarchization of spaces

6.1 Introduction

In the second half of the eighteenth century, Spain and Portugal sent demarcation parties to the Americas to establish the line between the dominions of the two monarchies. The boundary treaties entered into by both monarchies in 1750 and 1777, set forth that these parties were to be mixed, made up of Spanish and Portuguese commissioners who were to share instruments and agree on their demarcation tasks by drawing up joint maps, although it was not clear that both crowns effectively aspired to the success of their commissioners (Tejerina 1996, p. 34).

The creation of these mixed parties and the task entrusted to them of creating joint

Some of the ideas discussed in this paper were examined in: Rodríguez Arrillaga, Lucía, 'Milan and Paris in the Passo del Jacuy. The Circulation and Validation of Cartographic Images in a debate between Iberian Officials from Mid-Eighteenth Century Río de la Plata', paper presented at the VII Symposium of the International Society for the History of the Map, organised by the Max Planck Institute for the History of Science (MPIWG), Berlin, July 2023.

L. Rodríguez Arrillaga (✉)
Department of American History, Universidad de la República, Montevideo, Uruguay
e-mail: luciarodriguezarrillaga@gmail.com

J. García-Álvarez and A. Reyes Novaes (eds.), *Mapping the Contested Line*, Historical Geography and Geosciences, https://doi.org/10.1007/978-3-032-29889-8_6

maps clearly show that, establishing boundaries between the dominions of two monarchies in the course of the eighteenth century, required have maps that involved moving around the boundary spaces and taking measurements on the ground (García-Álvarez 2019, p. 18).

The mixed boundary commissions established in the eighteenth century in Europe and America had significant impact on the ways of conceptualising the spaces over which the monarchies wanted to exercise their exclusive sovereignty (Nordman 2007, pp. 162–167; Rodríguez-Arrillaga and Garrido-Pimenta 2021). The work of these parties contributed to the passage from 'a vision of space as a series of unique and differentiated places' typical of jurisdictional monarchies, to a vision of space closer to 'a geometric and homogeneous abstraction' (García-Álvarez and Puyo 2015) which would be the form of modern nation states, and which would later also be reflected in a profound transformation in the conceptions of sovereignty. Between the mid-eighteenth and mid-nineteenth centuries, the idea of a monarch who was sovereign in his dominions shifted to the idea of a 'national' or 'state' sovereignty, exercised over a certain 'territory' that had to be delimited as clearly as possible in order to make it a space of exclusion for other sovereign powers (Goldman 2014).

This process was not concluded with the demarcation parties, which did not complete their operations in the Americas. In fact, a definitive solution to the boundary issue was never reached, despite the two treaties entered into by Spain and Portugal for that purpose; it therefore became a problem inherited by the new political entities that emerged from the disintegration of both monarchies in South America.

The maps of the dividing line commissioned to the Spanish and Portuguese demarcators were not available for all European spaces, because both Iberian monarchies were slow to produce maps with all the available technology, in the style of the chart that France managed to produce in the mid-eighteenth century (Branch 2013). It was not until 1789 that Spain had a map of its coasts drawn up with astronomical data, but only in the late nineteenth century was it possible to draw up a map of similar characteristics for its entire territory. For its part, at the beginning of the nineteenth century, Portugal still did not have a map displaying the same characteristics, because a lack of resources at the end of the eighteenth century led to the abandonment of a triangulation project of all its peninsular territory. Thus, the Portuguese authorities decided to publish a Portuguese version of the compilation map printed in London by William Faden at the end of the eighteenth century (Valero 2015; Dias 2003; Moreira 2011).

The chapter focuses on the analysis of a debate between the Spanish and Portuguese commissioners sent to the Río de la Plata in South America after the signing of the Treaty of Madrid in 1750. Specifically, the commissioners discussed the location of the Ibicuy River, which pursuant to the dispositions of the 1750 treaty meant significant differences regarding the lands that corresponded to each monarchy. There are two dossiers concerning this dispute. While they include some overlapping interventions, each dossier presents some perspectives that do not appear in the other version. One of them is located in the *Biblioteca Nacional do Brasil* in Rio de Janeiro (BNB) (National Library of Brazil), and the other is held at the *Arquivo Histórico Ultramarino* in Lisbon (AHU) (Portuguese Overseas Historical Archive). These debates were common between Iberian officials, also during the demarcation tasks set forth by a new boundary treaty entered into in 1777, and were part of the reasons why a final agreement was never reached.

The transimperial nature of the Río de la Plata region contributes interesting information to understand the territorialisation process that the surface of the Earth was undergoing, and which acquired specific features in regions where two sovereign entities sought to create exclusive jurisdiction or sovereignty, acquired specific features (Maier 2016, pp. 83, 111). As a marginal space for both Iberian empires, one of the constitutive features of the region was the presence of subjects of other imperial powers in the Río de la Plata, such as the Dutch, English and French, and their business networks (Prado 2015).

In the context of these interimperial disputes, delimiting these spaces and producing knowledge about them was a way of appropriating them, and also a way of building a new legitimacy for such appropriation. With the signing of the treaty of 1750, Spain and Portugal abandoned *uti possidetis de iure*, which implied that the legitimacy of the occupation of spaces in the New World derived from the fact that they possessed those lands according to the law established by the papal donation and the Treaty of Tordesillas, and opted for *uti possidetis* de facto as the legitimating principle. This option implied recognising that it was the actual occupation of the territories that gave legitimacy to the European presence in America, and can be seen as a response to the constant criticism of the papal donation and the Treaty of Tordesillas from the other empires (Rodríguez-Arrillaga 2024).

Beyond the jurisdictional conflicts that the Iberian monarchies sought to resolve by means of treaties and the creation of demarcation parties, as was the case in other colonial spaces, the ultimate aim of the production of knowledge about the region was to dominate and obtain economic returns from these spaces, regardless of which of the two monarchies finally appropriated these attributes. In this vein, what was happening there was similar to the territorialisation processes that were taking place, practically at the same time, in other parts of the globe by other empires, albeit with different dynamics (Edney 1996).

This process involved social stakeholders from very diverse geographies, who brought with them specialised knowledge and a diversity of interests, which raises the question of the relationship between the production of cartographic information and the building of imperial powers (Anderson 1983; Black 1997). All the more so considering that since the fifteenth century various European powers 'took possession of the lands on paper before actually occupying them', so that 'maps anticipated empire' (Harley 2005, p. 85).

This was especially decisive for the nomadic peoples of the Río de la Plata, for whom the presence of the demarcators meant a slow but progressive impossibility of maintaining their own territorial dynamics. Because, despite its instability, in the medium term, the transit of Spanish and Portuguese demarcators and the material construction of the dividing line made it almost impossible for them to access these spaces (Erbig 2020; Herzog 2015).

The Río de la Plata presents an additional distinctive feature that makes it particularly valuable for examining how local cartographic culture was shaped by the demarcators and, conversely, how it influenced them. The prominent presence of the Jesuits from the early seventeenth century significantly impacted the social use of cartographic discourse in the region. The maps of the Río de la Plata region drafted on the basis of compilations made by the Jesuits, and which were used to illustrate books printed in European presses, were plentiful, especially after the eighteenth century (Furlong Cardiff 1936; Altic 2022; Asúa 2014). In addition, the Jesuits themselves and the Guaraní peoples living in their missions produced a very important set of larger scale, more detailed handwritten maps. Although the role of indigenous people in the authorship of these maps has been discussed (Barcelos 2013; Neuman and Barcelos 2022), it has been shown that they were used by them, as early as the seventeenth century, as 'notebooks' (Levinton 2010) in which to record information relevant to the recognition of land rights and to use them later 'as legal instruments and as devices of territorial memory' (Wilde and Takeda 2021, p. 611).

As discussed later in this paper, during the debate the demarcators discussed the reliability of these maps as sources of information, both with respect to the cartography printed in Europe based on the sources of the Jesuit fathers, and with respect to the manuscript maps produced by the missionary peoples. In some cases, they even acknowledged having used this cartography to create maps summarising their demarcation work, thus making said summary maps hybrid cartographic objects (Novaes 2020; Basset 1998). Nevertheless, for the missionary peoples, mastering the language of cartography was not enough to maintain access to the lands north of the Río de la Plata, where most of their lands were located, because the same process of linearisation of the frontier to which the Río de la Plata was subjected by both Iberian empires also meant the advance of Spanish and Portuguese landowners over those lands and over the cattle that had previously belonged to them.

In contrast to what happened with the parties created to demarcate the lands in the north of South America, in the region of the Orinoco and the Amazon rivers, where the commissioners never worked together, this did happen in the Río de la Plata (Bueno 2011; Lucena-Giraldo 1993, p. 175). In this case it is therefore possible to show the interimperial transfer of maps, forms of mapping and the circulation of people that made it possible.

A significant volume of handwritten maps was produced as part of the demarcation tasks. The historiography that traditionally analysed this corpus of maps placed special emphasis on the confrontations that justified the sending of these experts to map the region in order to put an end to the border disputes (Cortesão 2009, pp. 291–403). Despite the large number of disagreements between the parties, the cooperation and information sharing were inevitable. Therefore, this contribution focuses specifically on the information exchanges between the members of these parties, regardless of the challenges and tensions that these officials faced on the ground. The circulation of maps and plans among them and through the channels of the Iberian bureaucracies consolidated a transimperial 'cartographic common sense', which benefited the Spanish and Portuguese at first, but later also the French and the English.

In methodological terms, the debates and disagreements between the officials of both monarchies are considered to be an opportunity, as they show the circulation of cartography between the Iberian monarchies, which would eventually exceed them. An effective analysis requires a reconstruction of the trajectories of the contenders and the cartographic practices that they employed to defend their positions in the debate, because some commissioners served multiple empires, thereby acting as intermediaries in the transfer of knowledge (Rodríguez-Arrillaga 2024).

In turn, as these were jurisdictional disputes, the exchanges between the demarcators naturally had a singularising effect on the region that would take on a specific cartographic translation in these debates. Accordingly, when it came to determining the space over which their disputes were to be settled, the demarcators proposed a cartographic outline of the space which, in the case of the Río de la Plata, would become a fairly persistent cartographic outline.

On the other hand, from the epistolary exchanges between the commissioners of both empires and with their respective monarchies, it is possible to reconstruct the criteria of validation and hierarchisation applied in the production of knowledge about these spaces and their appropriation of local knowledge in the service of their respective empires.

In turn, his study is important because these debates bring to light the tension between two cartographic modes that was characteristic of cartography in the mid-eighteenth century (Edney 2019). One of these modes was based on the compilation of existing cartography and was mainly carried out in the cabinet, to produce mainly printed cartography; while the other mode was based on field surveys, although there was a later stage where all the information had to be compiled on the same map, which was mainly handwritten.

The possibility of measuring time in space, or in other words, the possibility of using precision instruments to account for the time spent on journeys, was what made it possible to produce modern cartographies such as those produced by the boundary demarcation parties. As noted by Mathew Edney, these new maps were progressively produced by civilian and military topographers and engineers who had accessed the specialised education of military academies and institutions and who relied on instruments to measure distances, angles and heights (Edney 2007; Bueno 2011).

However, these modes were not mutually exclusive, for as explained below, even with the best instruments, map making by compilation or by first-hand surveying alone was not the most common in the course of the eighteenth century. However, under the banner of greater accuracy and science, the latter mode would eventually prevail, becoming the cartographic mode par excellence throughout the nineteenth century. This process led to the emergence of a new term: 'cartography', which, not surprisingly, began to gain prominence in the early nineteenth century (van der Krogt 2015).

The chapter is structured around this introduction, a first section which introduces the discussion and the circumstances in which it developed, followed by a short section reconstructing some of the trajectories of those who took part in the controversy. The two sections below analyse the impact of the circulation of maps among those participating in the dispute on the configuration of specific cartographic outlines of the Río de la Plata, and on the validation of the cartographic modes involved in their production, as well as on the hierarchisation of the spaces of experience where these maps had been produced, namely Europe or locally, either by the communities of missionary peoples or by the members of the demarcation parties themselves.

Finally, the closing statements point out that, as in this debate about the linearisation of the frontier in the Río de la Plata, the movement of maps, forms of mapping and people matters if the criteria of validation and hierarchy that were in tension in the eighteenth century are to be reconstructed.

6.2 Discussion and Circumstances of the Cartography of Demarcation Lines

The objective of measuring and mapping the South American territories disputed between Spain and Portugal was to settle a jurisdictional conflict over these spaces. Thus, the demarcation parties were provided with the best instruments available at the time, which had already been used in previous scientific expeditions (Safier 2016; Puig-Samper 2017).

The maps were commissioned on the understanding that they would henceforth acquire evidentiary status; this was expressly stated in the text of the 1750 treaty. It can be argued that the evidentiary nature of maps required them to be made with the 'necessary observations' to fulfil this purpose, ensuring their accuracy and clarity for the future (Treaty of 1750, Article XI). In order to ensure the alleged accuracy of the maps, the instructions given to the demarcators by the Chief Commissioners, in May 1754, expressly stated that their production was the exclusive competence of the cosmographers 'because both Majesties recommend that the work they do should be not only accurate, but also conducive to the advancement of science' (Academia Real das Sciencias 1841, p. 38).

However, having the instruments available did not necessarily equate to the ability to use them, or at least not consistently, making the mandate for accuracy often difficult to fulfil (MacDonald and Withers 2015). The dilemma of this impossible accuracy accompanied the commissioners in their journeys through South America, and the tension between compilation and geometric measurements for the elaboration of their maps would be recurrent, sometimes simply to excuse themselves before their opponents, but other times, as part of their most private concerns and confessions.

The main challenge imposed on these demands for accuracy lay in the material dimensions of the task that they were entrusted with. The commissioners sent south in 1750 were to demarcate from Castillos Grandes to the mouth of the Jaurú River in the far north, organised into three subsets or parties (Cortesão 2009, p. 324). These were very large portions of the Americas, which lay approximately between sixteen and 35 degrees south latitude, the demarcation of which implied a distance of about 3000 kms in today's terms. However, given their unfamiliarity with much of the terrain they had to travel over, and as a result of the way they travelled at the time, the time it took to get from one place to another depended not only on the linear distances to be covered, but also on the characteristics of the terrain; this is why they used other types of measurements to calculate how long the journeys would take. They basically used the league, which was an itinerary measure that indicated the extension of land that could be covered in an hour's walk. It was a reference of distance typical of *ancien régime* societies, but which varied between different monarchies, and even between the different territories that made up the same monarchy.

Although the demarcators had begun their operations in the Río de la Plata in 1752, their work was slowed by the challenge posed by the Guaranitic War, which lasted from March 1753 to May 1758, and the second demarcation party, which was the last to complete its field operations, did so in 1760 (Cortesão 2009, p. 324). The Guaraní peoples resisted the implementation of the provisions of the Treaty of Madrid, also known as the Treaty of Permuta because of the mutual territorial concessions made by the two crowns, by force of arms. Regarding the Río de la Plata, the Portuguese undertook to hand over the Colonia del Sacramento, which they had founded in 1680 opposite Buenos Aires, on the strip located north of the Río de la Plata, while the Spanish agreed in exchange to hand over the territories where the missionary towns of the Upper Uruguay were located. Subsequently, the definition of the Guaraní peoples as subjects of one monarch or another would continue to change. The Treaty of 1761 restored the status quo prior to 1750, whereas the Treaty of 1777 ratified the status of the missionary peoples as vassals of the Spanish monarch. In 1801, however, part of them were occupied by the Portuguese army, and that was their situation when the wars of independence broke out.

As stated above, this paper analyses the dispute over the location of the Ibicuy River, which, according to the boundary treaty of 1750, had strong implications for the territories belonging to both monarchies. According to Ferreira, the persistence with which each side defended its position reveals the interests that mobilised them: the Portuguese wanted their dominion to extend into the interior of the continent and to reserve for themselves the usufruct of the 'fertile plains' of Rio Grande, while the Spanish wanted to avoid ceding more territory to the Portuguese, specifically the village of New San Miguel, whose location had resulted from the developments of the Guaranitic war (Ferreira 2001, p. 205; Zusman and Nunes 2022).

6.3 Circulation of People and Trajectories

The dispute was handled by the technical members of the parties (Ferreira 2001): Miguel Angelo de Blasco and José Custódio de Sá e Faria, on the Portuguese side, and Francisco de Arguedas and Francisco Milhau Maraval, on the Spanish side.

Before being appointed first commissioner of the second party, Arguedas had been a minister of the Court of Auditors, and after his appointment he spent nine years in America, three of which were spent in the field (Barreto 1973, vol. 1, p. 73). Milhau Maraval arrived in the Río de la Plata in 1751 at the age of 23, accompanying the leader of the Spanish parties, Marquis of Valdelirios. He stayed there until 1761 when he returned to Spain; between 1766 and 1771 his request to be sent back to Buenos Aires was granted, after which he returned definitively to Spain, where he died in 1805 (Barreto 1973, vol. 2, pp. 933–935).

The maps and reports produced by both Spanish officials, required by the monarchy at various times after their field experience in the Río de la Plata, played an important role during the negotiation of the new boundary treaty signed in 1777. On his return to Spain, together with Pedro de Cevallos and the Marquis of Valdelirios, Arguedas was a member of a Consultative Board attached to the Council of the Indies whose purpose was to prepare the negotiations for a new boundary treaty between Spain and Portugal in South America, which was finally signed in 1777. And one of the maps produced by Milhau Maraval after his demarcation work in the Río de la Plata was used by Cruz Cano y Olmedilla to delineate his map of 1777, although in a different projection (Smith 1966, pp. 55–56 and note 24; Almeida 2009, p. 82). But it was undoubtedly the two Portuguese representatives who had the most interesting trajectories, because having served several empires, not only maps but also forms of mapping circulated with them.

The trajectory of José Custodio de Sá e Faría is quite emblematic and an indicator that the Río de la Plata was a space of transimperial circulation. He arrived in the region in 1751, as a member of the Portuguese demarcation parties, and performed several tasks as a military engineer, in the region and in other Brazilian captaincies. In 1776, he was imprisoned by the expedition of Pedro de Cevallos, then first viceroy of the Río de la Plata, during the capture of Santa Catarina by Spanish troops. From then on, he settled in Montevideo and later in Buenos Aires, and worked for the Spanish Monarchy until his death in 1792 (Barreto 1973, vol. 1, pp. 486–510; Furlong 1948).

Miguel Ángel Blasco's trajectory is particularly interesting, his appointment was not an exception among Portuguese demarcators, because a significant number of them had previously been subjects of the Habsburg monarchy of Vienna and had carried out tasks in the demarcation of their dominions (Veres 2014. p. 82; Cortesão 2009, pp. 16–17; Ferreira 2001, p. 103, note 43). This shows that the linearisation of the border was not only a matter of concern for the Iberian empires regarding their American domains, but also for other empires with respect to their European domains. Furthermore, during the debate over the location of the Ibicuy River, Blasco was the most explicit in underlining the hierarchies for the validation of the maps cited in the dispute.

Miguel Ângelo Blasco was born in Genoa in 1697, and served the Habsburg monarchy in Vienna before moving to Lisbon to serve the Braganza monarchy, which was eagerly looking for military engineers to send to America. Together with his brother Dominico de Blasco, he had joined the Habsburg troops after the 1718 invasion of Sicily by Charles VI; they followed the Habsburgs and served in other provinces of the Monarchy for almost thirty years. In 1747, Blasco joined the newly created imperial brigades of engineers tasked with fortification and mapping work. From that position, he sought the mediation of his friends at the Court of Vienna to be promoted to the rank of colonel, but only managed to be conferred the rank of lieutenant colonel in 1745, and in order to attain that position he had to agree to transfer his services to the Portuguese crown (Veres 2014; Ferreira 2001, pp. 246–249).

Upon arriving in Lisbon, he was appointed Colonel of Infantry and Engineer in October 1750. He also secured the most favourable contract among the demarcators, receiving the highest compensation for his services. This position required him to relocate to Brazil, arriving in Rio de Janeiro in 1751(Golin 1998, pp. 180–181). Upon his return to Lisbon in 1758, he was promoted to the post of field marshal in 1763, and in 1769 he was appointed chief engineer of the Kingdom (Tavares 2000, p. 218).

From then on, he tried unsuccessfully to return to the service of Empress Maria Theresa (1740–1780), who finally decided not to hire him, because Blasco's having handled sensitive cartographic information could lead to a diplomatic conflict with Portugal. Blasco died in Lisbon in 1772. According to Veres, Blasco's interesting transimperial trajectory was possible because the cartographic demands of the different empires favoured the circulation of cartographers and their cartography, by hiring specialised personnel among the subjects of other imperial powers. From this derives Veres' important commitment to transcend imperial and national boundaries in order to study these trajectories, as this can also be a way of reconstructing the transfer of mapping technologies, in this case, between one imperial power and another (Veres 2014, p. 76).

6.4 Circulation of Maps and Cartographic Outlines

The debate was conducted in 1758 from Passo del Jacuy, a place where the demarcators were circumstantially located, partly because demarcation operations had been suspended as a result of the controversy. Its name originates from the fact that there was a crossing [*paso*] there, through which it was possible to traverse the river of the same name. It was from this lost place, which is now part of the State of Rio Grande do Sul in Brazil, and which did not appear on many of the maps that the demarcators carried in their travelling cabinets, that the cartography of South America produced in other regions was valued.

These two maps were included in the same Portuguese dossier which recorded the entire dispute, in order to illustrate the positions defended by each of the parties. Figure 6.1 shows a schematic version of the Map of the Two Courts, which was the foundation of the 1750 Treaty of Limits, where the Ibicuy River, the object of the dispute, did appear, but neither the Jacuy River nor the Passo del Jacuy were shown, which was where the dispute took place. The blue circle on this map circumscribes the space of the controversy. On the other hand, because the map in Fig. 6.2 uses a larger scale and is more detailed, it shows the Jacuy River and the Jacuy Pass, marked by the blue box on the map.

In the absence of a better alternative, the Map of the Two Courts had been imposed by the Portuguese during diplomatic negotiations and the Spanish eventually accepted it as a unit of meaning along with the treaty (Cortesão 2009, p. 251).

In general terms, with regard to the substantive issue, the location of the Ibicuy River, it can be said that Spanish geographers defended the existence of an Ibicuy River river that had not been considered during the signing of the Treaty, which had been done on the basis of the 'defective' cartography available until then. They relied on the cartography produced by the boundary and demarcation commissions themselves, as well as on the cartography of the Jesuit missions, which they used as a source of information for their own maps.

Conversely, the main argument used by the Portuguese demarcators regarding the location of the Ibicuy River was based on the cartography printed in Europe and on the Map of the Two Courts which had been based, in part, on that cartography.

The second map (Figs. 6.2 and 6.3) is a clear example of the circulation of cartography among the members of the demarcation parties, because it had originally been designed by the Spanish geographers to illustrate what for them was the 'correct' location of the Ibicuy River. And to further stress the controversy, when the Portuguese informed their court, they added inscriptions such as 'Ibicuy River put in question with this changed name, which was not demarcated in the General Map of the Two Courts, nor named in the Treaty', trying to reinforce that the location proposed by the Spaniards did not conform to the provisions of the Map of the Two Courts.

Both the demarcators, once in the field, and later historians, debated over which cartographic sources had been used to draw up the Map of the Two Courts, in order to determine the source of the 'errors' that made it of little use in contrasting the representation of space on the map with the 'configuration' of the terrain. The argument has finally been settled that the source for the delineation of the Paraguay River and the space between the Paraná and Uruguay rivers would have been the *Paraquariae provinciae Soc. Iesu* of 1726, but in the German edition of Mateus Seutter of 1732 (Ferreira 2007, p. 56). During the debate, Seutter's map was cited by demarcators on both sides, some to defend it and others to discredit it and use it to explain the 'problems' that the Map of the Two Courts had in using it as a guide for demarcation.

The disagreements over whether or not to use the Map of the Two Courts to guide the demarcations, which was part of this debate, have left us with some interesting pieces regarding the cartographic outline made by the demarcators, that is,

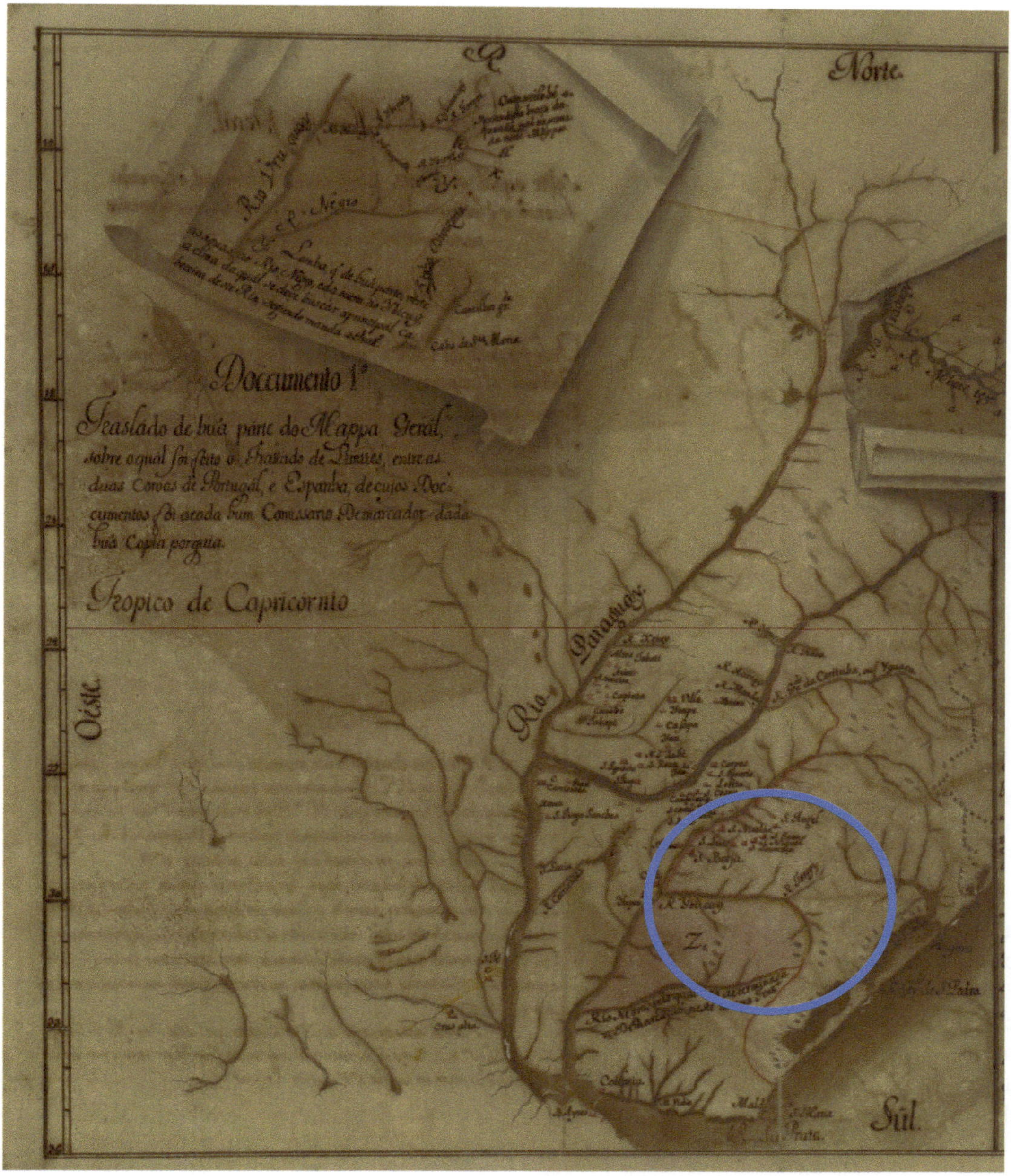

Fig. 6.1 Traslado de uma parte do Mapa Geral, [Partial transfer of the General Map], 1758. *Source* BNB, 049.02.024, f. 68

that of the territories between Asunción, Buenos Aires, Montevideo and Río Grande.

This is visible in Fig. 6.4, entered in the dossier that collected the whole dispute under the extensive title of 'Transcript of part of the General Map on which the Treaty of Limits between the two Crowns of Portugal and Spain was drawn up, a copy of which was given to each Demarcation Commissioner.' ['Traslado de uma parte do Mapa Geral sobre o qual foi feito o Tratado de Limites entre as duas coronas de Portugal e Espanha, de cujos Documentos foi

Fig. 6.2 Demonstration of the easiest and most exact way to ascertain the main origin of the Ibicuy and conclude the demarcation (*Demostración del modo más fácil y exacto para poder averiguar el origen principal del Ibicuy y concluir la demarcación*), 1758. *Source* BNB, 049.02.024, f. 8

a cada um Comissário Demarcador dada uma copia.'] (BNB, 049.02.024, f. 68).

From the title given to the map we can conclude that the demarcators called the Map of the Two Courts *'mapa Geral'* [general map], a name they also sometimes used to refer to the map used to summarise all the demarcation tasks performed by the corresponding party, so that it sometimes becomes confusing to determine which '*Geral*' map they are referring to in each case. The insert in the upper left quadrant exemplified in more detail the watercourses and 'lomba' that the treaty set forth as guides for the establishment of the line, the interpretation of which the commissioners of the respective monarchies disagreed with. While the insert in the upper right quadrant illustrates the outlet of the 'Mbote tey' river in Paraguay.

Although this rather schematic map represents the spaces between parallels nine and 36 of south latitude, it clearly singles out the area between the Tropic of Capricorn in the north and the Río de la Plata in the south. This cartographic outline was similar to that present in other maps of the same period, such as, for example, the 'Map of the Missions of the Society of Jesus by the Paraná and Uruguay Rivers according to the most modern measurements of latitude and longitude', compiled in the Río de la Plata by Jesuit priest José Quiroga in 1749, and published in Rome in 1753. This was a persistent cartographic outline for the Río de la Plata spaces, as it is also present in the maps

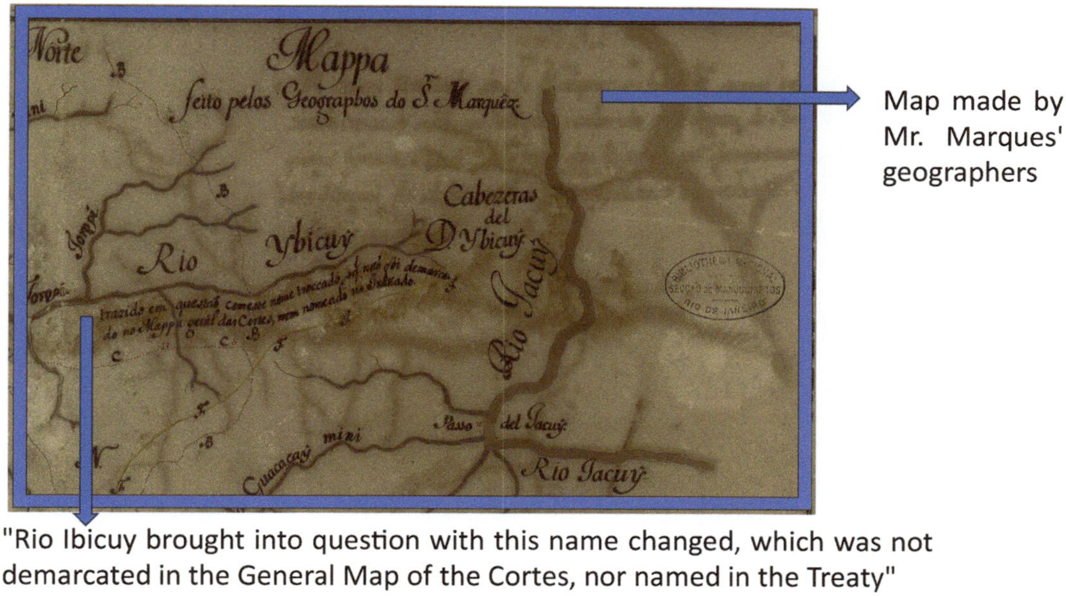

Fig. 6.3 Detail of Demonstration of the easiest and most exact way to ascertain the main origin of the Ibicuy River and conclude the demarcation, 1758. *Source* BNB, 049.02.024, f. 8

that summarised the demarcation work done by the subsequent parties created under the treaty of 1777. Like with Quiroga's map, this map was primarily illustrative of the other reports included in the dossier, yet it did not extend to the coast of Brazil beyond the Laguna de los Patos. Its interest lies in the fact that it allows us to trace the process of sedimentation of the new cartographic outline of the Río de la Plata that began to take shape in the mid-eighteenth century, as part of the jurisdictional dispute between the two monarchies over these spaces (Rodríguez-Arrillaga 2024).

6.5 Circulation of Maps, Cartographic Modes and Hierarchies

The maps referenced in the dispute over the location of the Ibicuy River were numerous, and although it is not always possible to ascertain the precise origins of each map in the possession of the commissioner citing it, it is clear that these maps circulated widely. The intention of this paper is not to review each case in which the demarcators acknowledged exchanging maps, but rather to present a few examples illustrating how these maps were employed in a biased manner to support the position of each side.

Table 6.1 has been compiled to summarise the maps cited by Spanish and Portuguese geographers, with a cross indicating the respective citer. Locally produced maps are marked in red, while those of which the demarcators exchanged copies are marked with a blue arrow, as it is this circulation of maps that helps to explain the configuration of a cartographic common sense about the Río de la Plata. All the maps produced in Europe and referenced in the controversy, highlighted in black in the table, were printed, with the exception of the Map of the Two Courts, which was the only handwritten one.

The reference to maps that had been produced in cabinets located miles away from the spaces represented in them is a clear indication of the dilemma that the demarcators faced. Because they were tasked with mapping territories they had not explored before, printed maps played a crucial role in representing these unfamiliar spaces. It is possible to note, however, that these

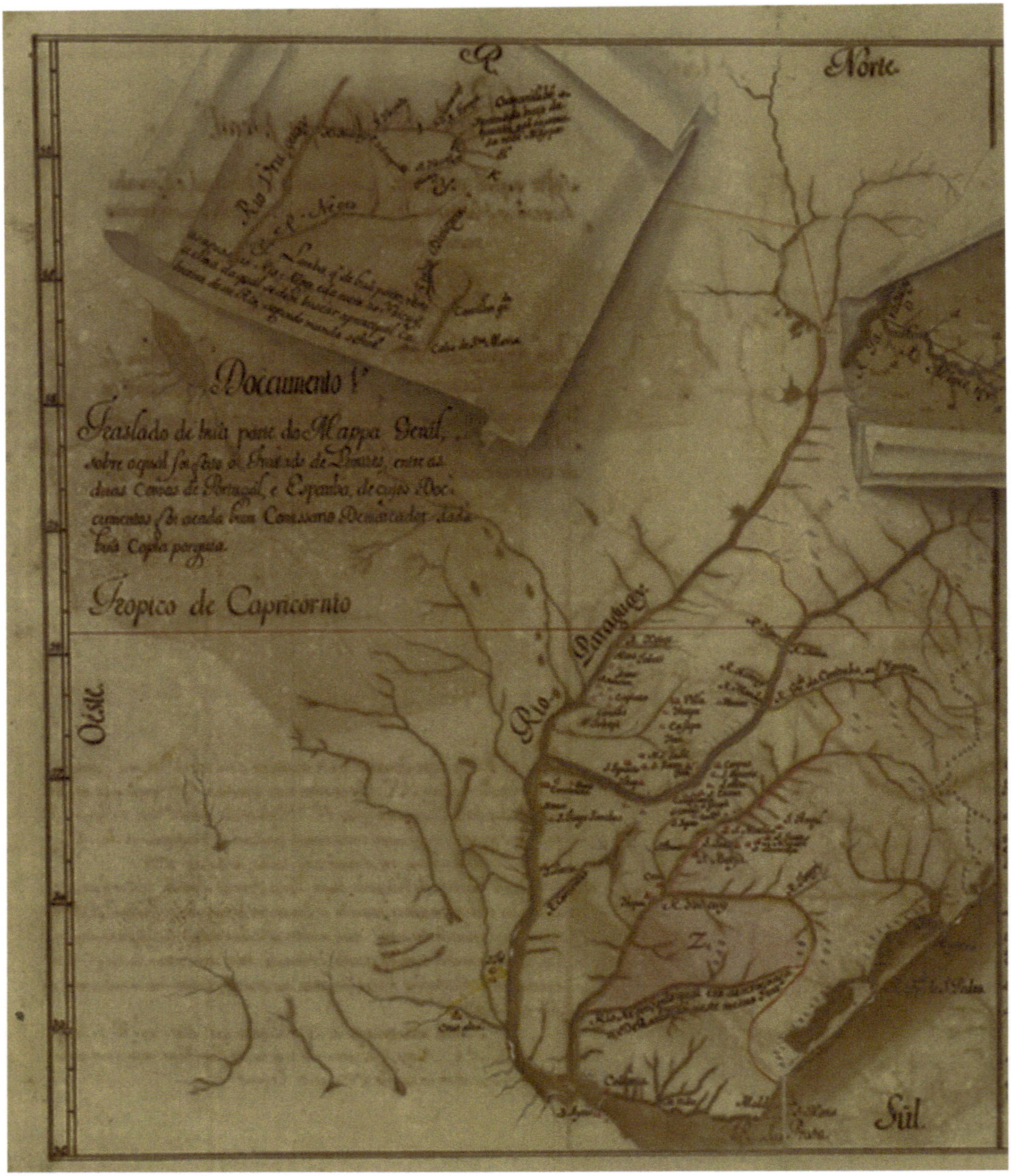

Fig. 6.4 Transcript of part of the Mapa Geral or map of the two courts, [Traslado de uma parte do Mapa Geral [o de las Cortes]]. *Source* BNB, 049.02.024, f. 68

maps were used in a quite biased manner, and how they validated their positions.

In one of his interventions in the debate, Blasco acknowledged that he had sent different maps to support his position regarding the location of the Ibicuy River. Among them he acknowledged having sent a copy of Father Enis' (Furtado 2023). In this case, the dispositions of this priest and the natives were well received by Blasco, who also acknowledged that he was sending Matteus Seutter's map of 1732, which

Table 6.1 Maps cited by Spanish and Portuguese commissioners during the dispute over the Ibicuy river

Lista de mapas mencionados en la controversia	Españoles	Portugueses
"el mapa general o de las Cortes"	x	x
Copia del "Mappa Original das Missões feito pelo Padre Enis"	x →	x
"Missions quas Provincia Jesu Paraguarica ex colitas flumina Parana et Uruguay ex nationes Guaranica accurate delineata à quodam e jusdem Missionario veterano", 1744	x →	
"Padre Josephus Quiroga societatis Jesu Propris observationibus de pinxit anno 1751 y por oral della propia rate pellimus undas", 1754	x	
"el del célebre Guillermo d'el Isle del año 1703"	x	
"el pequeño del teniente coronel José Custodio de Sá y Faria"	x	←
"los mapas jesuitas de los Archivos de la Candelaria y San Javier", "el del Padre Quiroga", "los del Padre Cardiel y Padre Ovando"	x	
"los Planos que mis compañeros y yo hemos levantado en los terrenos por donde hemos andado."	x	←
"los planos que levantó la 3ª Partida, así el común de la Demarcación como los Particulares de la Provincia del Paraguay"	x	
"Paraguaria Provinciae Soc. Jesu cum adjacentibs novíssima descriptio post itierate perigrinationes et plures Observationes Patrum.... accuratissime delineata & emendata", 1732		x
"todos os Mapas antigos e modernos até o ano 48", en que "o famoso Messieur de Anville publicou esta América Meridional"		x

Source Own elaboration based on the interventions of the Spanish and Portuguese commissioners in the dispute. Handwritten maps produced in the Río de la Plata are shown in red; the arrows show the map exchanges referred to in the interventions of each of the parties. The direction of the arrow shows who received the relevant map (BNB, 049.02.024)

had been one of the main sources for the Map of the Two Courts, as stated above:

> 'which, being based on the ancient documents of which some geography of this southern America is known, it is clear that until now there was no news in the world of any other Ybicuy, other than the one that the contracting Sovereigns ordered to be demarcated. The same justifies all the geographical maps that will be printed everywhere.' (BNB, doc. 1, f. 3)

On the other hand, Spanish geographer Milbau Maraval strongly questioned the quality of this 1732 map, considering that it contained significant errors in longitude, which had naturally been transferred to the Map of the Two Courts, on which it was based. He also acknowledged that the sources for his maps were highly diverse and, insofar as they were based on field survey operations, they directly contradicted both the Map of the Two Courts and the 1732 map.

> 'The materials I gathered for the formation of my large map are various printed and handwritten maps, diaries and astronomical observations made by different subjects and the plans that my companions and I have drawn up in the areas where we have walked.' ['Los materiales que junté para la formación de mi mapa grande son varios mapas impresos y manuscritos, diarios y observaciones Astronómicas hechas por diferentes sujetos y los Planos que mis compañeros y yo hemos levantado en los terrenos por donde hemos andado.'] (BNB, doc. 2, f. 32)

In his compilation, he was careful to compile only what he felt was 'certain', and preferably what had been surveyed at first hand by the members of the parties:

> 'I only included in m[y map] for certain the information surveyed by a member of this commission, whether Spanish or Portuguese, and the rest, which I took from the maps, is subject to the corrections that the demarcation parties might make in view of the terrain. [Solo puse en el mío por seguro aquello que se ha levantado por alguno, ya Español, ya

Portugués, de los individuos destinados a esta comisión y lo demás tomado de los mapas la dejo sujeto a las correcciones que hiciesen a la vista del Terreno las Partidas Demarcadoras.]' (BNB, doc. 2, f. 33)

As for the validation and hierarchisation of the knowledge produced by Baquianos and natives, as opposed to the European cabinet cartographers, they also had discordant positions. Let us look at some examples of the validation of these cartographic images and of the hierarchisation of the spaces of experience in which these images were produced or used.

The experience in the Jesuit archives during the expedition and the Guaranitic war was valued very differently by the Spaniards and the Portuguese. On the one hand, the Spanish geographer granted on several occasions that he had been guided by the information contained in these maps, which validated the location of the Ibicuy River that he defended:

> 'In the trip that I made to the towns on the other side of the Uruguay with my Commissioner Don Francisco Arguedas, by order of Your Lordship, I recognised the maps that he took from the archives of Candelaria and San Javier, by which I learned part of the course of this river, and in the latter town Father Lorenzo Ovando gave me a plan that the Fathers gave him, sending him in the years 40 to 50 with a party of a thousand Indians to recognise the fields of this side and repel the Charrúas who were pursuing the estancias of these towns. The same Father told me that he had found this map, which shows the Estancias in different colours, was in accordance with the terrain, and that he had added some Ports as he went along, and I have also used it for some of my additions. [En el viaje que de orden de Vuestra Señoría hice a los Pueblos de la otra banda del Uruguay con mi Comisario Don Francisco Arguedas reconocí los mapas que sacó de los Archivos de la Candelaria y San Javier por los que supe parte del curso de este Rio y en este último pueblo me dio el Padre Lorenzo Ovando un Plano que le dieron los Padres enviándole por los años de 40 a 50 con una partida de mil Indios a reconocer los campos de esta banda y rechazar a los charrúas que perseguían las estancias de estos Pueblos.'] (BNB, doc. 2, f. 32)

Blasco, on the other hand, believed that 'The natives, raised without any science or doctrine did marvellous things in making the imperfect maps they made' but 'because they did not have any knowledge of geography' their maps were not reliable and he even discredited them to the point of considering them 'useless papers':

> 'I did not have the opportunity to enter the Archives of the Fathers of Candelaria, as the two gentlemen did, but since I know that the aforementioned maps and manuscripts are of the same quality as those we found in Santo Angelo, where our General Headquarters was, I know the price that can be made of them, most of which contradict each other by hitting each other blindly, as I can also show by some that were left to me by carelessness in the discarding of useless papers. ['Eu não tive ocasião de entrar no Arquivo dos Padres da Candelaria, como tiveram os dois Senhores, mas como sei que os citados Mapas e manuscritos são da mesma qualidade dos que achamos no de Santo Angelo, onde foi o nosso Quartel Geral, bem sei o preço que se pode fazer deles, em cujas a maior parte se contradizem dando-se-se entre si pancadas de cego, como posso ainda mostrar por alguns que por descuido me ficaram no descartar de papeis inúteis.']' (BNB, doc. 3, f. 40)

However, as stated above, Blasco had been a clear advocate of Seutter's 1732 map, which had also been printed in Europe based on the Jesuit surveys. However, in this instance, the map in question did serve to substantiate his position on the location of the Ibicuy River.

During the controversy, Blasco was an outspoken advocate of the European cabinet cartographers, of which group he otherwise considered himself a member. Following his argument, we will now present another example of how the spaces of production and circulation of cartographic knowledge about South America were hierarchically organised.

While drawing on his social status in Milan as an authoritative criterion in favour of his position, Blasco also acknowledged that he used these printed maps to take proportionate account of the tasks entrusted to him with regard to spaces that he deemed difficult to represent from a distance, while he was still in Europe:

> 'I saw many of them in Milan when I went to say goodbye to the Fathers of the Company and to my friends Marquez Trivulsi, Simonetta, Lucini etc., all of them curious about Geography. I did not bother to take note of the authors or any other circumstance, either because I could not foresee these doubts or because the aforementioned Fathers and with them the aforementioned noblemen told me that my task should be in the

divisions within the jurisdiction of the Bishops; but I always noticed that in all the maps of this part of America there was no more than a single Ybicuy, whose headwaters were close to those of the Rio Negro, so that when I came across the famous chart of South America by the famous Mr. Anville, printed in the year 48, which I still keep, the boundary treaties that Count Borri, a nobleman who was very learned in all the arts and sciences and geography, showed me, I was able to draw a line on it, using the treaty as guide, not to see how the demarcation was going, but how far the road that I had to take on this expedition extended.' [Eu não cuidei então tomar nota dos Autores nem de outra circunstância assim porque não podia eu prever estas duvidas quanto porque os sobreditos Padres e com eles os ditos Fidalgos me diziam que a minha incumbência devia ser nas divisões da jurisdição dos Bispados; mas sempre reparei que em todos os Mapas desta parte d'América não havia mais do que um só Ybicuy, o qual tinha as suas cabeceiras pegadas as do Rio Negro, de sorte que achando-me me as mãos a celebre Carta desta América Meridional do famoso Mr. De Anville impressa no ano 48, que ainda conservo, os tratados de Limites que me fez ver o Conde Borri Fidalgo muito instruído en todas as Artes, e Ciências e coisas Geográficas, eu pude tirar sobre la uma Linha com a guia do dito tratado não para ver como corria a demarcação mas quanto se estendia o caminho que eu devia fazer nesta Expedição.'] (BNB, doc. 3, f. 40)

As stated above, one of the main arguments of Spanish geographers against European printed cartography was what they considered to be a lack of accuracy regarding the longitudes. Blasco considered himself to be one of these cabinet cartographers, given his previous work for the Habsburgs in Vienna, He therefore argued that it was impossible to make astronomical measurements of the entire space to be mapped, using the case of Italy as an example:

> 'In this regard, who could fail to recognise Tholomeo, Magini, Galileo, and one (if I remember correctly) named German Kraister as great men in all the sciences and mathematical faculties? These famous authors each made their own geographical chart of Italy, against which the above-mentioned Mr. De Anville (who is still alive) in his *Itinerere d' Italie* makes a significant change of about three degrees of longitude and I cannot remember how much of latitude; whose errors will also be demonstrated by me when God is served, that I finish once and for all this Holy Demarcation that I may give the last stroke to my Map of Italy, which I left already in good condition arranged in 48 sheets of large paper.' ['A este proposito quem será que não reconheça Tholomeo, Magini, Galileo, e um tal (se bem me lembro) nomeado Kraister Alemão por homens grandes em todas as Ciências e faculdades Matemáticas? Fizeram estes famosos Autores cada um de per si uma Carta Geográfica da Itália, contra as quais o acima mencionado Mr. De Anville (que inda vive) no seu Itinerere d'Italie lhe faz enorme reforma de perto de três grãos de Longitude e não me Lembro quanto de Latitude cujos erros se hão de ver também demostrados por mim quando Deus seja servido, que acabe de uma vez esta Santa Demarcação que eu possa dar a ultima mão ao meu Mappa de Itália, que deixei já em bom estado disposto em 48 folhas de papel grande'] (BNB, doc. 3, f. 40)

6.6 Closing Remarks: Movement Matters

The circulation of maps and individuals is crucial for understanding how criteria of validation and hierarchy are constructed and employed in a manner that reflects self-interest, particularly in relation to the spaces involved in the production of cartographic images. From the analysis of the correspondence and the reconstruction of the paths followed by the printed and handwritten maps referred to in this debate, it was possible to determine which criteria of authority and hierarchy were employed regarding the reliability of their sources of knowledge.

This paper has shown some examples of transimperial circulation of maps, with the excuse of locating the Passo del Jacuy, and from the list of maps cited by the disputing parties. It also attempts to show some forms of validation of cartographic images and hierarchisation of the spaces of experience in which these images were produced. It has also proven the way in which the contenders used their bias to mobilise this hierarchy, based on which for some, d'Anville's Chart printed in Paris in 1748 or what the 'Geography curious' of Milan had to say about territories an ocean away, had more value than the maps produced locally by 'natives raised without any science or doctrine' or the maps themselves produced by the officials of the respective monarchies.

Based on the evidence presented thus far, it can be inferred that although the Iberian kingdoms favoured handwritten cartography, in order to minimise the circulation of geographical information about their American dominions, it is nevertheless true that the information circulated through various channels that are better known today. This requires that we qualify the effectiveness of this secrecy and to think about how many of the cartographic initiatives projected within the framework of the interimperial disputes collaborated unintentionally in the sedimentation of cartographic knowledge and cartographic outlines that always circulated beyond the borders of these empires and contributed to the sedimentation of a transimperial 'cartographic common sense' about the Río de la Plata, which firstly benefited the Spanish and Portuguese, of course, but later also the French and the English. It is no coincidence that in the first decade of the nineteenth century, a new cartographic image of the Río de la Plata was publicised in Europe that capitalised on the knowledge produced by Spain and Portugal in the context of their border conflicts. Examples include the Map of Part of the Viceroyalty of Buenos Ayres, printed by Aaron Arrowsmith in London in November 1806, or the map of the Gouvernement de Buenos Ayres printed in Paris in 1809 as one of the maps in the Atlas that accompanied Félix de Azara's Voyages dans l'Amerique Meridionale (Rodríguez-Arrillaga 2024).

On the other hand, Spanish and Portuguese engineers used the cartography printed outside Madrid and Portugal as input for their survey operations in the field, and for the subsequent drafting of the maps that summarised these operations. The realisation of this reversibility between manuscript and printed cartography helps to understand what a formidable challenge it was for European cabinet cartographers in the eighteenth century to rely on field-produced information when drawing their printed maps. On the other hand, it also shows the challenges faced by the engineers tasked with representing spaces that they had to move through and measure without having thorough knowledge of them.

In addition to using these maps as a criterion of authority to defend interested positions in their confrontations over demarcation tasks, it is possible to assume that printed maps could be of use to them because they could help them to imagine spaces that were immeasurable for them in practical terms, especially until they had maps summarising the partial plans produced in the course of their demarcation tasks.

Even for those who practised the new mapping techniques developed in the eighteenth century, spatial measurement and precision taken to the extreme were regarded as an impossible task. Even if these officials had all the necessary tools, they could not establish the line and the maps resulting from their joint work based solely on the patterns of the new cartographic mode that was taking over, based on field surveys, because it would take time they did not have.

Archival Sources

Tratado de Madrid, 1750.

AHU Arquivo Histórico Ultramarino, Lisbon. 'Ofício do [coronel engenheiro], Miguel Ângelo de Blasco ao [secretário de estado do Reino, Sebastião José de Carvalho e Melo], enviando documentos sobre as questões suscitadas pelo [comissário principal espanhol da Demarcação dos Limites da América Meridional], marquês de Valdelírios, [D. Gaspar de Munive León Garabito Tello y Espinosa], e seus geógrafos no acerca da continuação dos trabalhos da 1a Partida da Demarcação dos Limites, parada em Santa Tecla por causa dos índios rebelados." ('Dossier from [Colonel Engineer] Miguel Ângelo de Blasco to the [Secretary of State of the Kingdom, Sebastião José de Carvalho e Melo], sending documents on the issues raised by the [Spanish Chief Commissioner for the Demarcation of the Boundaries of South America], the Marquis of Valdelírios, [D. Gaspar de Munive León Garabito Tello y Espinosa], and his geographers regarding the continuation of the work of the 1st demarcation party, stopped in Santa Tecla because of the rebellious natives.') Jacuy, 10 October 1758, AHU-CU [Conselho Ultramarino] Brasil-Limites 059, Cx. 2, D. 142.

BNB Biblioteca Nacional do Brasil, Rio de Janeiro. 'Notta dos papeis, que vão no masso junto, feitos pelo Coronel de Blasco, por ordem do Exelentisimo Marichal de Campo General e Comissrio Plenipotenciario Gomes Freire de Andrada, em resposta às questoens novamente sucitadas do Senhor Marquez de Valde Lirios, e seus geographos, a respeito da demarcação de Primra Partida, no acto de continuala neste anno de 1758' (Note of the papers that are in the enclosed mass, made by Colonel de Blasco, by order of His Excellency Field Marshal General and Plenipotentiary Commissioner Gomes Freire de Andrada, in response to the questions raised again by Mr. Marquis de Valde Lirios, and his geographers, regarding the demarcation of Primera Partida, in the act of continuing it in this year of 1758) (BNB, 049,02,024).

BNB, doc. 1: 'Judgement or opinion of the said Colonel [Blasco] on a letter of 9th April and a small map sent from the Marquez de ValdeLirios'. [Juízo ou parecer do dito Coronel [Blasco] sobre uma carta de 9 de Abril e pequeno mappa remetido do Senhor Marquez de ValdeLirios.] 19 April 1758, (BNB, 049,02,024).

BNB, doc. 2: Letter from the Geographer Don Francisco Milbau Maraval, dated 20 May, in which he gives the promised reply from Mr Marquez.', (BNB, 049,02,024).

BNB, doc. 3: Post scripto do dito Coronel [Blasco] em 16 de junho em resposta ao dito Geografo, Desde Passo do Jacuy, 16 de junho de 1758, (BNB, 049,02,024).

References

Academia Real das Sciências (1841) Collecção de noticias para a historia e geografia das Nações Ultramarinas que vivem nos dominios portuguezes, ou lhes são visinhas, vol 7

Altic M (2022) Encounters in the new world. Jesuit cartography of the Americas. The University of Chicago Press, Chicago

Anderson B (1983) Comunidades imaginadas. Reflexiones sobre el origen y la difusión del nacionalismo. Fondo de Cultura Económica, México

Barcelos A (2013) O Mergulho no Seculum: exploração, conquista e organização espacial jesuítica na América espanhola colonial. Ed. Animal

Barreto A (1973) Bibliografía sul-riograndense. A contribuição portuguesa e estrangeira para o conhecimento e a integração do Rio Grande do Sul, 2 vols. Conselho Federal de Cultura, Rio de Janeiro

Basset T (1998) Indigenous mapmaking in intertropical Africa. In Woodward D, Lewis MG (eds) The history of cartography, vol 2, book 3. University of Chicago Press, Chicago, pp 24–48

Black J (1997) Maps and politics. The University of Chicago Press, Chicago

Branch J (2013) The cartographic state. Maps, territory, and the origins of sovereignty. Cambridge University Press, Cambridge

Bueno B (2011) Desenho e desígnio. O Brasil dos Engenheiros Militares: 1500–1822. Edusp

Cortesão J (2009) História do Brasil nos velhos mapas. Imprensa Nacional, Casa da Moeda, (1st edition 1965)

Dias MH (2003) As explorações geográficas dos finais de Setecentos e a grande aventura da Carta Geral do Reino de Portugal. Revista da Faculdade de Letras, Geografia, I(XIX):383–396

de Almeida AF (2009) O Mapa Geográfico de América Meridional, de Juan de la Cruz Cano y Olmedilla. Anais Do Museu Paulista 17(2):79–89

de Asúa M (2014) Science in the vanished Arcadia: knowledge of nature in the Jesuit missions of Paraguay and Rio de la Plata. Brill, Leiden

Edney M (1996) Mapping an empire: the geographical construction of British India, 1765–1843. University of Chicago Press, Chicago

Edney M (ed) (2007) Cartography in the European enlightenment, part 1, the history of cartography project. Chicago University Press, Chicago

Edney M (2019) Cartography. The ideal and its history. The University of Chicago Press, Chicago

Erbig JA (2020) Where caciques and mapmakers met. Border making in eighteenth-century South America. The University of North Carolina Press

Ferreira MC (2001) O Tratado de Madrid e o Brasil Meridional. Os trabalhos demarcadores das partidas do sul e a sua produção cartográfica (1749–1761). Comissão Nacional para as Comemorações dos Descobrimentos Portugueses

Ferreira MC (2007) O Mapa das cortes e o Tratado de Madrid: a cartografía a serviço da diplomacia. Varia Historia 23(37):53–69

Furlong G (1936) Cartografía Jesuítica del Río de la Plata, Vol. I. Publicaciones del Instituto de Investigaciones Históricas Facultad de Filosofía y Letras, Number LXXI, Talleres S.A. Casa Jacobo Peuser

Furlong G (1948) José Custodio de Sá y Faría, Ingeniero, Arquitecto y Cartógrafo colonial, 1710–1792. Anales Del Instituto De Arte Americano e Investigaciones Estéticas 1:11–74

Furtado JF (2023) A Guarani war-era manuscript map between Indians, Jesuits and Luso-Brazilian Colonizers. Paper presented at the VII symposium of the international society for the history of the map, organised by the Max Planck Institute for the History of Science (MPIWG), Berlin

García-Álvarez J, Puyo J-Y (2015) La aportación geográfica y cartográfica de las Comisiones de Límites luso-franco-españolas (siglos XVIII y XIX). Terra Brasilis, Nova Série 6

García-Álvarez J, Puyo J-Y (2019) Los primeros intentos de delimitación moderna de la frontera pirenaica: la Comisión franco-española Caro-Ornano y su legado cartográfico (1784–1792). Revista De Historiografía 30:15–44

Goldman N (ed) (2014) Soberanía. In: Fernández-Sebastián J (dir.) Diccionario político y social del mundo iberoamericano La era de las revoluciones, 1750–1850, Vol. X. Centro de Estudios Políticos y Constitucionales/Universidad del PaísVasco

Golin T (1998) A guerra guaranítica. O levante indígena que desafiou Portugal e Espanha. Editora da Universidade, Passo Fundo, UPF Editora

Harley JB (2005) La nueva naturaleza de los mapas. Ensayos sobre la historia de la cartografía, Fondo de Cultura Económica

Herzog T (2015) Frontiers of possession: Spain and Portugal in Europe and the Americas. Harvard University Press

Levinton N (2010) La micro-región: Espacio y tiempo en la cartografía producida por la interacción jesuítico-guaraní. Anu Estud Am 67(2):577–604

Lucena-Giraldo M (1993) Laboratorio tropical: La Expedición de límites al Orinoco (1750–1767). Monte Ávila University and Consejo Superior de Investigaciones Científicas, Madrid

MacDonald F, Withers CWJ (2015) Geography, technology and instruments of exploration. Routledge

Maier C (2016) Once within borders. Territories of power, wealth, and belonging since 1500. Harvard University Press

Moreira LM (2011) O Mapa Corografico dos Reinos de Portugal e Algarves Copiado do Ingles de W. Faden (1809): construir a imagem de um país. Paper presented at the IV Simpósio LusoBrasileiro de Cartografia Histórica, Porto

Neuman E, Barcelos A (2022) A razão gráfica missioneira. Escrita e cartografia indígena nas reduções da América colonial. Martins Editora

Nordman D (2007) Boundary disputes and cartography. In: Edney M (ed) Cartography in the European enlightenment, part 1, The history of cartography project. Chicago University Press, pp 162–167

Novaes AR (2020) Mapping cross-cultural exchange: Jaime Cortesão's dialogues and documents on the role of indigenous knowledge in Brazilian exploration. In Schelhaas B et al (eds) Decolonising and internationalising geography. Essays in the history of contested science. Springer, pp 1–16

Prado F (2015) Edge of empire. Atlantic networks and revolution in Bourbon Río de la Plata. University of California Press

Puig-Samper MA (2017) La medida de América: de la observación métrica ilustrada española al empirismo razonado humboldtiano. Hist Mex 67(2):907–963

Rodríguez-Arrillaga L (2024) Fisiografías apasionadas. Territorio y patriotismo en el cruce de los imperios ibéricos: el Río de la Plata 1750–1810. PhD dissertation, University of Sao Paulo, Brazil

Rodríguez-Arrillaga L, Pimenta JP (2021) Conceptos espaciales en el Río de la Plata a finales del siglo XVIII y comienzos del XIX: tierras, terrenos, territorios y dominios. Ariadna histórica. Lenguajes, Conceptos, Metáforas 10:51–87

Safier N, (2016) La medición del Nuevo Mundo. La ciencia de la Ilustración y América del Sur. Fundación Jorge Juan and Marcial Pons Historia, Madrid

Smith TR (1966) Cruz Cano's map of South America, Madrid, 1775: its creation, adversities and rehabilitation. Imago Mundi 20:49–66

Tavares, AL (2000) A Engenharia militar portuguesa na construção do Brasil. Editora Biblioteca do Exército

Tejerina M (1996) La lucha entre España y Portugal por la ocupación del espacio: una valoración alternativa del Tratado de San Ildefonso de 1777. J Hist 135:31–40

Valero A (2015) Arte y Ciencia: El Atlas Marítimo de España de Vicente Tofiño. Imafronte 24:73–102

Van der Krogt P (2015) The origin of the word 'cartography'. e-Perimetron 10(3):124–142

Veres MV (2014) Unravelling a trans-imperial career: Michel Angelo Blasco's mapmaking abilities in the service of Vienna and Lisbon. Itinerario 38:75–100

Wilde G, Takeda K (2021) Tecnologías de la memoria: Mapas y padrones en la configuración del territorio guaraní de las misiones. Hisp Am Hist Rev 101(4):597–627

Zusman P, Nunes S (2022) Borders on the paper, borders on the ground: boundary commissions under the Treaty of Madrid and their role in the reinvention of the Rio de la Plata frontier (1752–1759). In: García-Álvarez J, Puente-Lozano P (eds) Beneath the lines: borders and boundary-making from the 18th to the 20th century. Springer, Cham, pp 93–108

Knowledge Mobilities and Archival Assemblages: Mapping Practices in the Brazilian-Bolivian Borders

7

André Reyes Novaes

Abstract

This chapter examines border mapping in South America during the early twentieth century, arguing that maps acquire distinct meanings and functions depending on their archival contexts and spaces of circulation. It analyses the archives of boundary commissions preserved in the collections of the Royal Geographical Society (UK) and the Museu Histórico e Diplomático do Itamaraty (Brazil), addressing the mobility of knowledge and the formation of archival assemblages as ways to challenge colonialist narratives in the history of boundary-making. Focusing on the practices of mapping and demarcation along the Verde River—on the Brazilian-Bolivian frontier between 1908 and 1909—the chapter moves beyond national frameworks to uncover multiple narratives and standpoints in the production of geographical knowledge. Examining the archives of two military engineers who served respectively as subchief and chief of the Bolivian and Brazilian boundary commissions, Percy Harrison Fawcett and José Cândido Guillobel, the paper highlights the potential of combining materials across languages, institutions and locations to construct a transnational perspective on border-mapping practices.

Keywords

Border mapping · Archives · Knowledge mobility · Brazil–Bolivia border · Percy Fawcett · Cândido Guillobel

7.1 Introduction

This chapter examines border mapping in South America during the early twentieth century, drawing on two research approaches that have gained increasing traction among historical geographers and historians of cartography. The first emphasises knowledge mobility, focusing on the circulation of people, artefacts, images (and maps) to develop a critical and transnational understanding of knowledge production (Zusman 2015; Jöns et al. 2017). The second regards archives as world-making, combining documents dispersed across institutions and regions to challenge Eurocentric narratives of exploration and mapping (Hodder et al. 2021, Raj 2007).

This work has been supported by the Conselho Nacional de Desenvolvimento Tecnologico (CNPq, Brazil, Number 308958/ 2022–5) and the Spanish State Research Agency under Grant Number PID2020-114088 GB-I00/ AEI/10.13039/501100011033

A. R. Novaes (✉)
Universidade Do Estado Do Rio de Janeiro, Rio de Janeiro, Brazil
e-mail: andre.reyes.novaes@uerj.br

J. García-Álvarez and A. Reyes Novaes (eds.), *Mapping the Contested Line*, Historical Geography and Geosciences, https://doi.org/10.1007/978-3-032-29889-8_7

These two methodological concerns—circulation and archival combination—have guided my research over the past few years, particularly during archival visits to consult maps, diaries and reports produced by the boundary commissions that surveyed the Brazil–Bolivia border in the early twentieth century. Working between the Museu Histórico e Diplomático do Itamaraty (MHDI) archives in Rio de Janeiro and the Royal Geographical Society Archives (ARGS) in London, I encountered overlapping yet discontinuous figures and narratives—encounters that prompted reflections on how knowledge is produced and circulates at the border. I argue that these incomplete and fragmentary histories help embody and situate the practices of border mapping beyond national frameworks and geopolitical reasoning, revealing multiple standpoints in the production of geographical knowledge (Raj 2007).

The relevance of combining archives in border-mapping research became particularly evident when I found two maps depicting the Rio Verde, along the Brazil–Bolivia border in 1908. These maps serve as an entry point into the individuals, narratives and mobilities explored throughout this chapter (Fig. 7.1). Both trace the course of the Rio Guaporé through the old town of Mato Grosso to its confluence with the Rio Verde, which then ascends towards the rugged terrain of the Serra de Ricardo Franco, where its source lies. A comparison reveals that the second map is a monochrome reproduction of the first—a copy of an original

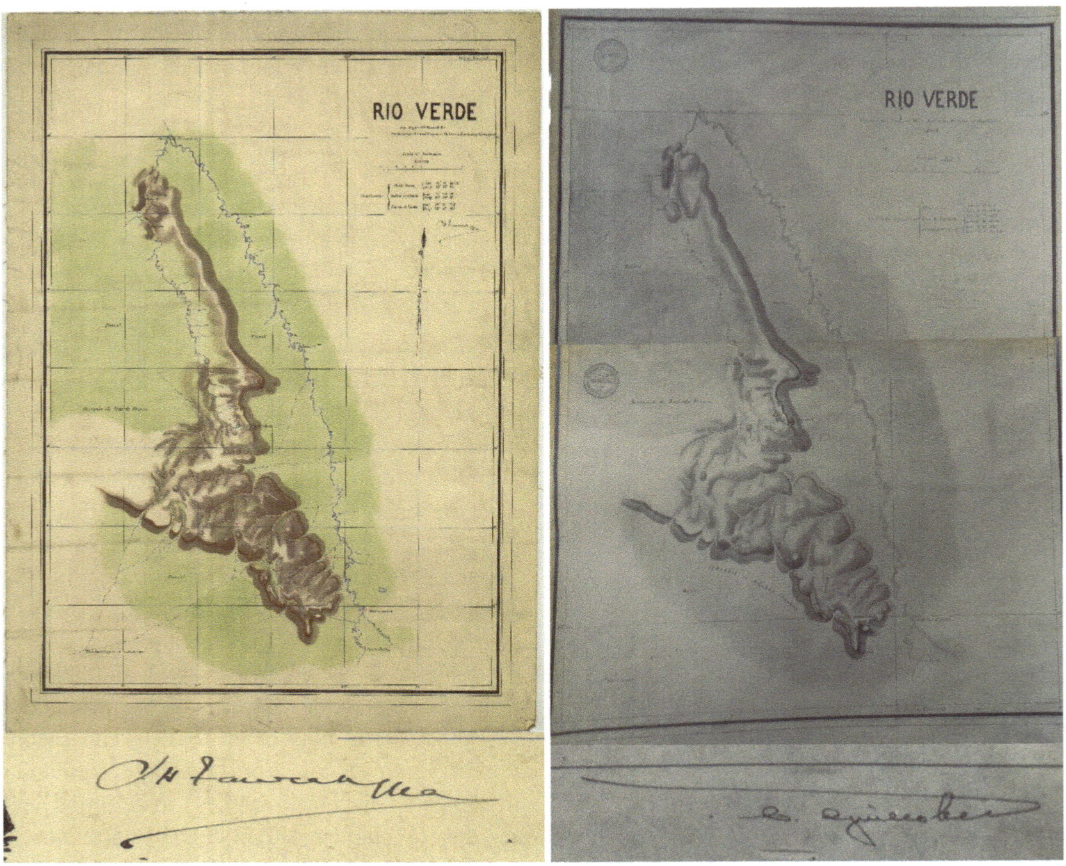

Fig. 7.1 Percy Harrison Fawcet. Rio Verde, 1908. *Source* Royal Geographical Society. José Cândido Guillobel. Rio Verde, 1908. *Source* MHDI and Mapoteca Histórica do Itamaraty (MHI)

manuscript rendered in colour. Beyond differences in medium, however, the maps diverge significantly in authorship: the coloured map is signed by Colonel Percy Harrison Fawcett, a British military engineer serving the Bolivian Boundary Commission, while the copy bears the signature of Cândido Guillobel, head of the Brazilian Boundary Commission.

Percy Harrison Fawcett, the author of the first map, was born in Torquay, England, in 1867. After serving in Ceylon and southern Africa, he was invited by the Royal Geographical Society to undertake border mapping in Bolivia. He arrived in La Paz in 1906 and continued working for the Bolivian government until 1912, producing maps of Amazonian border rivers such as the Rio Acre and the Rio Heath (Novaes 2020a, b). Yet the river that posed the greatest challenge to his work was the Rio Verde, which he explored in 1908 and 1909 (Fawcett 1909).

The author of the second map, José Cândido Guillobel, was nearly twenty years older than Fawcett. Born in Rio de Janeiro in 1843, Guillobel was a naval officer who played a prominent role in mapping and negotiating Brazil's borders. He took part in the Joint Commission for the Survey and Exploration of the Missions Territory (1887) and contributed to the negotiations leading to the Treaty of Petrópolis with Bolivia in 1903. Following this agreement, he was appointed head of the Brazilian Boundary Commission responsible for demarcating the frontier with Bolivia. From Corumbá, Guillobel coordinated operations across various border zones. Although he did not personally explore the Rio Verde, he assigned a Brazilian officer to lead the survey there in 1909.

The fact that Guillobel's team relied on a copy of Fawcett's map could easily lend itself to a simplistic—and all too common—narrative: one in which Fawcett appears as the pioneering explorer who 'discovered' and accurately mapped the Serra de Ricardo Franco and the source of the Rio Verde. Taking Fawcett's own words at face value, we might conclude that his work made it possible for every mile of the river to be 'accurately mapped', thereby correcting the supposed 'guesswork' of previous commissions (Fawcett 2001 [1953], p. 125).

However, Brazilian sources complicate this narrative, revealing controversies, power dynamics and competing interests within the mapping process. In correspondence addressing 'the problems with the English soldiers', Guillobel accused the British officers of being more concerned with exploring natural resources than with conducting rigorous cartographic work (Guillobel 1907). He also criticised the poor condition of the instruments used by the British, claiming that this rendered 'any accurate survey based on astronomical observations untenable' (Guillobel 1907).

The maps of the Rio Verde, therefore, carry different meanings depending on their archival contexts. As images are never isolated, a map's significance emerges through the networks of documents, texts and other images that surround it. Combining archives across languages, institutions and locations allows us to forge new connections between border documents, contributing to a transnational perspective on mapping practices.

The chapter unfolds in three sections. The first strengthens the methodological argument by presenting recent scholarship on circulation and mobility that bridges archives across linguistic and institutional boundaries to challenge localist, nationalist and colonialist perspectives. The second situates Brazil–Bolivia border negotiations within the context of the rubber economy and its geopolitical implications. The third examines the 1908–1909 expeditions along the Rio Verde, tracing divergences and convergences in the records preserved at the Royal Geographical Society and the Itamaraty archives.

7.2 Knowledge Mobility and Archival Assemblage

In an article published in Terra Brasilis in 2015, Professor Perla Zusman makes a statement that continues to reverberate across debates in historical and critical geography: 'it is not only stillness that constitutes geographies' (Zusman

2015). With this phrase, Zusman opens a theoretical and methodological space for thinking geography as a discipline fundamentally marked by movement, exchange and circulation. Her argument goes beyond the metaphorical use of mobility to describe cultural or intellectual contact; it positions mobility as a constitutive dimension of geographical practice and of the very production of spatial knowledge. By emphasising circulation as an analytical category, Zusman invites us to rethink how geographical ideas travel, are transformed and take root in different contexts. In doing so, she dismantles the notion that geography can be understood solely within the confines of the nation-state or through the authority of European scientific traditions. For her, the key lies in overcoming methodological nationalism—the assumption that intellectual and scientific developments are bounded by national contexts—and recognising instead that knowledge has always been produced through transnational networks, collaborations and conflicts (Quijano 2000; Mignolo 2003a, b). This shift carries profound epistemological implications: it enables the critique of entrenched hierarchies that position certain regions as producers of theory and others as mere recipients or imitators.

It is worth recalling, however, that the attention to movement and circulation is not an entirely new theme within geography. As Zusman herself acknowledges, mobility has long been a recurrent concern—though often treated unevenly and through diverse conceptual vocabularies. Already in the nineteenth century, Friedrich Ratzel considered migration and displacement as driving forces in the evolution of human societies, developing an interest in Chinese migrations and their role in shaping the global diffusion of cultures and technologies (Ratzel 1876; Seemann 2012). Ratzel's diffusionist insights, as Moraes (1990) reminds us, helped lay the groundwork for thinking about the circulation of innovations as a spatial process. Later, geographers such as Jean Brunhes and Camille Vallaux explored the relationships between nomadism, circulation and territorial organisation (Vallaux and Brunhes 1921), emphasising that spatial life is constituted as much by movement as by settlement.

In the early twentieth century, Carl Sauer extended this line of thought by examining how domesticated plants, animals and agricultural techniques travelled through colonial networks, giving rise to new environmental and cultural landscapes (Sauer 1952). Similarly, Torsten Hägerstrand's diffusion models (1952) translated the idea of circulation into a quantitative framework, seeking to explain how innovations spread through time and space. Finally, Doreen Massey's relational conception of place (2008) reframed geography's understanding of space, arguing that places are not static containers but rather intersections of social and material flows. From these examples, one can see that geography has never been a discipline of immobility. On the contrary, it has always understood space as a dynamic and negotiated production, shaped by the ceaseless movement of people, goods, ideas and representations.

Despite these important precedents, it was only in the 1990s that geographers began to systematically investigate the circulation of ideas as an object of study in its own right. The publication of Edward Said's 'The World, the Text and the Critic' (1983, 2004, p. 303) marked a decisive turning point by suggesting that ideas 'travel from person to person, from situation to situation and from one period to another'. This simple but powerful observation prompted a generation of geographers and historians to attend to the trajectories, mediations and translations through which geographical knowledge circulates. In the Brazilian context, scholars began to examine the localisation of scientific and political discourses—what Machado (2000) called the study of 'ideas in place'. These works explored how concepts imported from Europe or North America were adapted to local debates about modernisation, nation-building and the management of tropical nature.

Such studies represented a crucial move beyond the paradigm of 'national schools' that had long dominated the historiography of geography. They revealed that geographical knowledge has always been produced in circulation, through the translation of concepts, the adaptation of methods and the appropriation of external ideas. Nevertheless, as Zusman (2015) points out, even these innovative works often retained a national framework of analysis. They tended to treat the nation-state as the natural unit of intellectual history, thereby overlooking the cross-border movements, transoceanic collaborations and interstate controversies that shape geographical practice (Wimmer and Schiller 2002).

Zusman's call to abandon methodological nationalism thus marks an important epistemic rupture. She proposes that we understand transnational spaces of knowledge production—arenas where scientists, explorers, diplomats and technicians engage in sustained interactions across frontiers. These spaces are not peripheral to the making of science; they are sites of innovation and experience, where new methods and interpretations emerge. Viewing border regions in this way also illuminates the central role of what might be termed peripheral science in the global production of knowledge. Much of what has been canonised as universal science was, in fact, generated through the work of local intermediaries, technicians and intellectuals operating in colonial or semi-colonial settings, later reclassified through the European locus of enunciation (Mignolo 2003a, b). As Pratt (1993) reminds us, scientific knowledge was forged in 'contact zones', where asymmetrical encounters produced hybrid epistemologies. To recognise this hybridity is to challenge the fiction of European exclusivity in the making of modern science.

These theoretical advances have encouraged scholars to seek 'hidden histories' and to give voice to 'intermediaries' and 'go-betweens'—figures who connected distant intellectual worlds and facilitated the material and epistemic exchanges that sustained global networks of knowledge (Driver and Jones 2009; Konishi et al. 2015). Among the most influential figures in this historiographical turn is Kapil Raj, whose research on 'connections, crossings and circulations' has reshaped our understanding of colonial science and its geographies (Craib 2009). In Relocating Modern Science (2006), Raj challenges the dominant narrative that portrays modern science as a linear, Eurocentric project diffused outward from metropolitan centres to passive peripheries. Instead, he demonstrates that science emerged through interaction—through negotiations, accommodations and exchanges that took place in specific historical and geographic contexts. His analysis of the mapping of India in the eighteenth century exemplifies this approach, showing how the production of colonial cartography depended on a complex ensemble of local knowledge, technical skill and intercultural collaboration.

Raj's critique directly engages with the work of Matthew Edney, who, in his early studies of the mapping of India, had highlighted the imposition of mathematical and colonial rationalities on local spatial practices. Edney described how European surveyors introduced a new, 'scientific' conception of space that contrasted with what they perceived as the 'mystical' or 'religious' spatialities of India (Raj 2007, p. 156). Raj, however, exposes the limitations of this interpretation by cross-reading archives from both England and India. His comparative archival strategy reveals that British mapping projects relied profoundly on local expertise. The British presence in India was thin, lacking the trained personnel needed for extensive surveying. Indeed, as Raj observes, around 1760—when the first major land surveys began—there was still no detailed, unified map of the British Isles themselves (Kain and Baigent 1992).

By examining Indian archives, such as those preserved in the National Archives of India, Raj demonstrates that local map-making traditions predated colonial interventions. Though no single map of the entire subcontinent existed, a rich corpus of regional maps already represented large parts of western and central India (Gole 1989; Phillimore 1952). Detailed cadastral maps depicted cultivated areas, routes of communication and irrigation systems (Raj 2007, p. 69). These maps employed locally meaningful units of measurement—such as maize cobs or bamboo rods—recorded in administrative memorials and reports. Instruments of Arabic manufacture, including astrolabes traded across Southeast Asia, were also widely used for celestial and terrestrial measurements (Gunther 1932).

When these tools and techniques intersected with European surveying practices, they generated hybrid cartographies that combined distinct epistemic traditions. Far from being mere by-products of imperial rationality, colonial maps were the outcome of negotiations, frictions and mutual adjustments between different knowledge systems (Raj 2007, p. 77).

Nevertheless, as Raj shows, when these hybrid artefacts travelled to Europe—through publication, exhibition and archiving—their complex origins were effaced. The map of India produced by Major James Rennell in 1782, for instance, became a symbol of British scientific achievement once it reached London. Upon awarding Rennell a medal in 1791, the president of the Royal Society proclaimed that 'England should be proud', as the mapping project positioned the nation as the 'Queen of scientific advancement' (Raj 2007, p. 78). This metropolitan appropriation exemplifies how archives and institutions at the centre re-signify and reify colonial hierarchies, transforming collaborative achievements into imperial triumphs. Unless researchers actively seek to combine multiple archives and question the silences embedded in metropolitan collections, the colonial logic of authorship and mastery continues to dominate the historical record.

The methodological lesson drawn from Raj's work is crucial: the circulation of knowledge is not a unidirectional transfer from centre to periphery, but a process of negotiation and transformation that produces new scientific practices and epistemologies. This approach, indebted to connected history and entangled history frameworks (Subrahmanyam 1997), has profoundly influenced recent scholarship on the contributions of indigenous, vernacular and intermediary actors to the making of modern science. While this body of work is essential—and one to which I have sought to contribute (Novaes 2020a, b)—the present paper turns towards a different but related set of concerns: the geopolitical controversies and practices that shaped the boundary commissions responsible for mapping the frontier between Brazil and Bolivia.

Building on these reflections about knowledge mobility and archival assemblage, my investigation of boundary commission archives raised several guiding questions. How might the combination of dispersed archival sources illuminate the power relations embedded in map-making? To what extent can attention to the circulation of people, instruments and visual materials foster a more critical and transnational understanding of border cartography in South America? How did the maps produced along the frontier circulate, and who were their intended audiences? And finally, how were their meanings transformed as they moved through different institutional and political contexts—whether displayed, published, or stored?

In the sections that follow, I offer an exploratory discussion of these questions, focusing on the cartographic controversies that unfolded along the Brazil–Bolivia border, with particular emphasis on the mapping of the River Verde.

7.3 Rubber Exploitation and Growing Tensions on the Brazil-Bolivia Border

Why did the Rio Verde become an important river in border demarcation? The controversy surrounding the source of this river had accompanied diplomatic agreements between Brazil and Bolivia, and to understand the exploration of this river in the early twentieth century by the boundary commissions, it is important to contextualise negotiations between these South American countries in previous centuries. The first border treaties in the region are connected to the dispute of colonial powers derived from the Spanish and Portuguese empires. The Treaty of Tordesillas, from 1494, considered the 'unknown' area as part of the Spanish empire, but Luso-Brazilian colonial actors advanced westward during 250 years along the river roads of the Amazon and Paraguay systems (Fifer 1966). This advance via the Amazon, Madeira,

Guapore and Paraguay rivers was recognised in 1777, by the Treaty of San Ildefonso, pushing the border until the Javary River. However, the area remains largely unknown to central colonial powers, and understanding the sources and confluences of the region's rivers is a constant factor throughout the border negotiation process (Vergara 2010).

In the nineteenth century, with the independence of South American countries, new negotiations were established to define the region's boundaries. In 1834, the Bolivian government sent its representative to Rio de Janeiro to conduct negotiations aiming to establish the border demarcation based on the Treaty of Santo Ildefonso (1777). The Brazilian Empire ultimately rejected the proposal, seeking to prioritise the forts constructed after the 1750 treaty. Negotiations would resume in greater depth only in 1867, when the first agreement between the countries was approved under the title 'Treaty of Friendship, Limits, Navigation, Commerce and Extradition', known as the 'Treaty of Ayacucho'. The treaty establishes the oblique Munoz-Netto line, which connects the unknown area between Madeira and Javary. Brazil expands its territory towards the Amazon and agrees to divide some lagoons, such as Cáceres, Madioré, Gaíba and Uberaba. According to many authors, the primary objective in these negotiations was to ensure Bolivia's neutrality in the war against Paraguay (Goes Filho 1999), but the growing exploitation of rubber gave renewed importance to the region (Resor 1977).

The interest of the Brazilian Empire in rubber commercialisation started before the war, as shown by decrees such as the 1853 Amazon Steam Navigation Company. The introduction of steamships on the Amazon allowed a new time scale in the circulation of products, because the distance from Belém to Manaus, for instance, dropped from months to an eight-day journey (Resor 1977). The commercialisation of rubber growth exponentially during the next decades, and the high prices of the 1860s attracted the attention of the British Foreign Office. Clements Markham, the head of the office and a member of the Royal Geographical Society, instructed the British consul in Para, James Hay, to find someone who would be willing to collect some seeds for shipment to London (Resor 1977, p. 343). It was in this context that Henry A. Wickham, a young planter at Santarém, sent 70,000 seeds to London using a brand-new British steamer, the Amazonas, built especially for the Brazil trade.

The historical consequences of this appropriation and circulation of seeds are well known, resulting in the decline and collapse of rubber production in South America. After being packaged and domesticated at the Royal Botanic Gardens, Kew, these seeds were transported to Southeast Asia, specifically to Malaysia, where production was organised more efficiently to break the South American monopoly. Nevertheless, while the British Empire sought to establish a new production site, large profits were made from exploring South American forests, which intensified border disputes in the Amazon.

In this context, the lines between Brazil and Bolivia became more disputed and less generic and geometrical. The treaty from 1867, for instance, renewed a previous proposition to use the source of the River Verde as a border reference. In a memoir written a decade earlier, diplomat Duarte da Ponte Ribeiro had already commented on the different proposals for defining these boundaries, contrasting the geometric and straight lines of the 1777 Treaty of São Idelfonso, which defined a line up to the confluence of the Sararé and Guaporé rivers (represented in purple), with the lines using the source of the Verde River as a reference (represented in red) (Fig. 7.2). Border negotiation and demarcation began to be done inch by inch, even though in parts of the land that were still unknown.

Demarcation work under the 1867 treaty only began in 1870, following the end of the War of the Triple Alliance (Brazil, Uruguay and Argentina) against Paraguay. By 1878, much of the southern section of the border had been explored and demarcated, with subsequent conferences between the governments ratifying the new boundary lines (Vergara 2010). However, even during the exploration, mapping

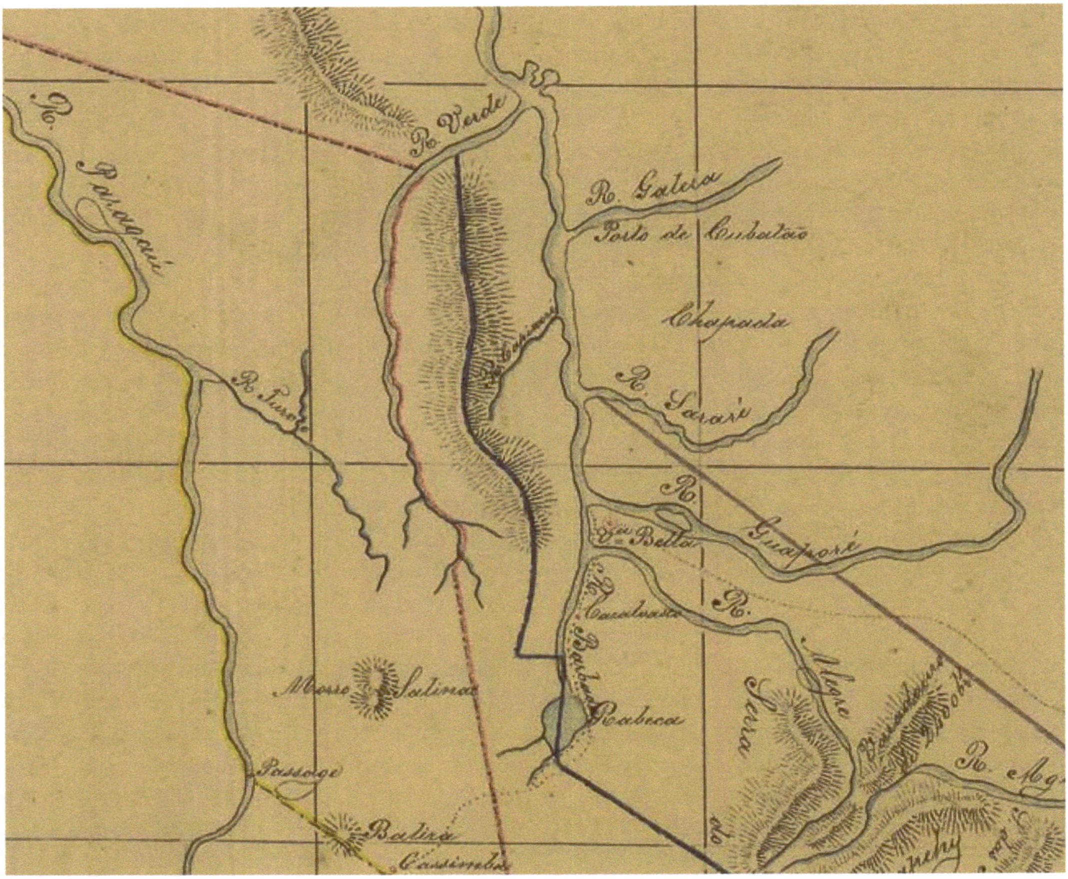

Fig. 7.2 Map accompanying Duarte da Ponte Ribeiro's 'Memória sobre os limites e negociações do Brasil com as repúblicas do Peru, Bolívia e Paraguai' (1842). *Source* MHDI and MHI

and placement of boundary markers, numerous doubts and uncertainties surrounded the work of the boundary commissions. One such controversy involved the source of the Verde River, which the 1874 boundary commission had erroneously identified. Seeking to avoid a deeper incursion into the Ricardo Franco Mountain range, the commission mistakenly identified an entirely different river—the Turvo—as the Verde. These uncertainties would shape later explorations and fuel subsequent territorial disputes.

Tensions surrounding rubber exploitation also remained high in the region. Between 1890 and 1913, the Upper Amazon became the world's most important centre for rubber production, thanks to its monopoly and the exceptional quality of raw materials from Hevea brasiliensis and Hevea lutea (Resor 1977). The rapid growth of the rubber economy attracted thousands of migrant workers from Northeastern Brazil during the 1880s, intensifying pressures along the Amazonian borders. Poorly defined boundaries and limited knowledge of the headwaters of major rivers—such as the Javari and the Verde—enabled substantial Brazilian expansion into the Amazon rainforest. By 1899, the population of the Acre region had reached approximately 60,000 (Fifer 1966), and local settlers declared independence, posing a direct challenge to Bolivian territorial claims.

To secure its hold over the contested area, Bolivia granted a vast triangular tract of roughly 75,000 square miles to an Anglo-American enterprise, the Bolivian Syndicate, for rubber extraction. Several British and American syndicates already controlled large concessions in Bolivian territory, building a rubber economy heavily dependent on foreign investment. However, this particular concession along the border provoked intense diplomatic and military tensions. The dispute was ultimately resolved through the Treaty of Petrópolis, signed in 1903, negotiated by Brazilian Foreign Minister José Maria da Silva Paranhos, the Baron of Rio Branco and Bolivian Minister Cláudio Pinilla. Carlos Guillobel, who was also performing in the Petropolis negotiation table, was later appointed head of the Brazilian Boundary Commission with Bolivia. In a famous photograph from the diplomatic meetings, which was taken at Baron Rio Branco's house, Guillobel was the third figure from left to right (Fig. 7.3). Rio Branco and Claudio Pinilla occupied the centre of the photograph, demonstrating their main role in negotiating the new treaty in which Bolivia ceded the territory of Acre to Brazil. Brazil, in turn, surrendered lands in the Paraguay River basin, including Bahia Negra and the Abunã Triangle. Brazilian diplomacy also agreed to construct the Madeira–Mamoré Railway and paid Bolivia £2 million to compensate the Anglo-American Bolivian Syndicate.

After the agreement, new demarcation processes were triggered and, to address the

Fig. 7.3 Photography from Petropolis Treaty diplomats and negotiators. The Baron Rio Branco and the Bolivian Ministry Claudio Pinilla are in the centre, and José Candido Guillobel is the third from left to right. *Source* MHDI and MHI

lack of qualified personnel, the Bolivian government requested the Royal Geographical Society of London to appoint British military engineers to lead its boundary commissions. One of these officers was Percy Fawcett, who was invited for a conversation with the Society's president in 1905. According to the British military, the president invited him to his office and asked, 'Do you know anything about Bolivia?' (Fawcett 2001 [1953], p. 18). Describing the Northern frontier between Bolivia, Peru and Brazil, the president pointed on a map to 'an enormous area of tropical and plains not yet by any means completely explored', a region of rubber country where 'the potential wealth is enormous' (Fawcett 2001 [1953], p. 18). The president explained that since the late nineteenth century, the high rubber prices attracted investments and political attention to a region previously neglected by national states, arousing the question of 'what territory belongs to whom'. Fawcett accepted the challenge, arriving in Bolivia in 1906 to assist in a demarcation process that served both Bolivian and British rubber interests. The circulation of these two men, Guillobel and Fawcett, led them to cross paths in the Brazilian-Bolivian boundary, raising different controversies in the mapping of the River Verde. The River Verde is still very poorly understood by the central powers and is still an important source to delimit the boundary marks. It was not by chance that the foreign ministers Rio Branco (Brazil) and Pinilla (Bolivia) issued joint instructions calling for renewed exploration of the river in 1907 (Ganzert 1934). The ministers' call stimulated two expeditions to the River in 1908, by Fawcett and the Bolivian Boundary commission, and in 1909, by the two boundary commissions together. The controversies of these explorations, considering the combination of archives from the Royal Geographical Society and the Itamaraty Historical Archives, are discussed in the previous section. It is through the reports and maps signed or received by both figures that I aim to interrogate a narrative which overstates the role of British military engineers in the mapping of South American borders in the early twentieth century.

7.4 Controversies and the Circulation of Maps in the Mapping of the Rio Verde, 1908–1909

The first meeting between Fawcett and Guillobel took place in May 1908, when the Commissions convened in Corumbá and began mapping the surrounding lakes. According to Fawcett, 'the Brazilian Boundary Commission received us on board with great ceremony. The commander of the garrison was with them, and champagne was served in the ship's saloon' (Fawcett 2001 [1953], p. 114). In his letters and reports, Guillobel also refers to exchanges of information with British engineers, who provided data on the longitudes used by the Bolivian commission (Guillobel 1909). It is unclear whether Fawcett shared any information with Guillobel aboard that vessel anchored on the Rio Paraguay, or whether Guillobel, in turn, offered Fawcett any data or mapping plans. Perhaps the two men simply drank champagne together politely.

In any case, the encounter between Fawcett and Guillobel was short-lived. Fawcett soon delegated the 'tedious routine' of survey work in the vicinity of Corumbá to a French engineer, choosing instead to set off in search of 'the true source and course of the mysterious Verde' (Fawcet 1909, p. 183; 1910). More inclined towards discovery than what he described as the 'boring work' of detailed mapping, Fawcett sought to 'clear up the embarrassing doubts' surrounding the Verde's course (Fawcett 2001, p. 116). He organised an expedition accompanied by an English colleague and 'half a dozen peons', explaining that 'the Brazilian Commission declined to supply a representative, and politely considered us mad' (Fawcett 1909, p. 183). This first expedition marked the beginning of a long-standing controversy between Fawcett and Guillobel—one whose meaning and significance are presented in strikingly different

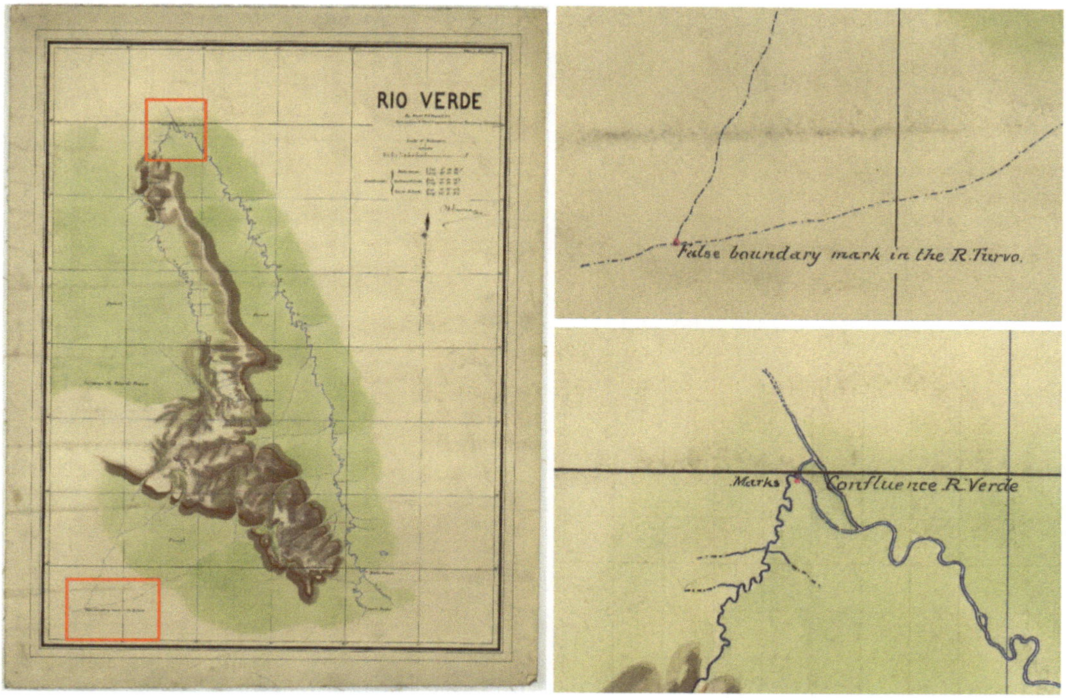

Fig. 7.4 Source of the Rio Verde by Major P. H. Fawcett 1909. 1:30 000. 22 × 26. Details from the false mark at the River Turvo and the confluence between Guaporé and the Verde. *Source* ARGS [Collection of maps by Maj. P. H. Fawcett F.R.G.S, 1907–1910] Printed with permission

ways in the two archival collections that document their work.

Fawcett departed from Corumbá with his group, heading into Mato Grosso to reach the Rio Guaporé. They intended to correct the 'false boundary mark' placed in the Rio Turvo by the previous commission, and Fawcett made a point of including the wrong border marker on his map (Fig. 7.4). The inclusion of the mistaken point was fundamental to Fawcett's narrative, which sought to convince his audience of the pioneering nature of his exploration and mapping. They navigated along the Guaporé and entered the Verde at its confluence, where an old border marker was located. The group travelled up the Verde and began ascending the Serra de Ricardo Franco, where mobility difficulties began with greater intensity.

The Ricardo Franco Mountain was difficult to climb due to the steep terrain and numerous waterfalls. According to Fawcett, 'the peons could carry no more than their rifles and ammunition, their hammocks, and a theodolite' (Fawcett 1909, p. 183). Although he frequently referenced his instruments, Fawcett also often remarked on the difficulty of using them. The theodolite and chronometer saw little use during the journey, and his field notebook reveals the varied methods he employed to determine longitude, including 'chronometric difference', 'azimuth calculation' and 'triangulation' (Fig. 7.5). This constant shift in methods used to calculate longitude suggests that Fawcett struggled to adapt scientific practices to a hostile environment. His scientific work with the English theodolite could not be replicated under the conditions of the Serra de Ricardo Franco, demonstrating how the circulation of instruments and techniques needs to be adapted and calibrated to different environmental conditions.

Reading Fawcett's accounts, it becomes clear that the boundary work was carried out by

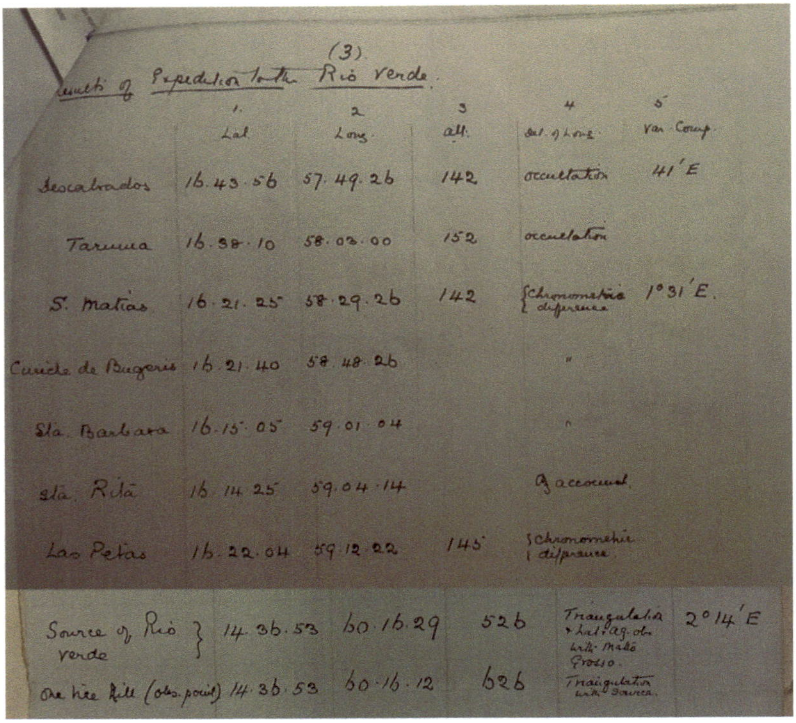

Fig. 7.5 Field notebook from Percy Harrison Fawcett, Rio Verde. The notebook documents the different methods used by Fawcett to calculate the latitude and longitude of various points in River Verde. *Source* ARGS [Fawcett Collection, 1907–1910]. Printed with Permission

a hungry team, hauling heavy instruments under extremely precarious conditions (Fig. 7.6). Nonetheless, Fawcett was received with enthusiasm when he returned to London to present his findings at the Royal Geographical Society in 1909. Clearly influenced by a culture of exploration that celebrated sacrifice and the endurance of hardship (Driver 2001), Fawcett used the lecture room's projector to display a photograph of his team at the source of the Rio Verde. The lack of food and the gruelling ascent, he claimed, had reduced the group to 'a mere collection of bones' (Fawcett 1909, p. 183). It seems unlikely that the team conducted a systematic survey of the terrain, but despite the difficulties, Fawcett praised the 'tremendous effort' made to triangulate the source of the river with the town of Mato Grosso (Fawcett 2001, p. 122). According to his narrative, the expedition was a success, and all prior theories about the Rio Verde were dismissed. If the river marked the boundary, Fawcett wrote, 'it was essential to map it with complete accuracy, or [we could] ruin the whole purpose of the expedition' (Fawcett 2001, p. 121).

For an audience unfamiliar with the terrain and listening to Fawcett from the comfort of a London lecture hall, the explorer's maps may have seemed relatively convincing. In addition to his 1909 handwritten map of the Rio Verde, Fawcett also produced a smaller-scale map representing the entire northwestern state of Matto Grosso. On this map, the Rio Verde appears quite generically, with virtually no representation of the relief on Ricardo Franco Mountain (Fig. 7.7). Despite its limitations, this was the map reproduced in Fawcett's article in the Geographical Journal (Fawcett 1910). The Rio Verde appeared schematically and with few details, but an epic narrative of overcoming obstacles accompanied the map, and its reception recognised the British explorer's efforts to map unknown places never before visited.

Among the Brazilian commission, however, Fawcett's maps are far from being trustworthy or pioneering, and the Brazilian commission

Fig. 7.6 Photograph by Percy Harrison Fawcett (who is the taller of the two central figures) Rio Verde. This picture was shown by Fawcett in his presentation at the Royal Geographical Society in 1909. *Source* P. H. Fawcett, Exploration Fawcett. 2001, p. 142.

criticised the fact that he had undertaken an initial expedition without the participation of both commissions. In light of this mistrust and the practical need to place boundary markers on the ground, Guillobel requested a new joint expedition to the headwaters of the Rio Verde in 1909. Guillobel's archives at the Itamaraty reveal a different history of frontier exploration, in which the British soldier was far from being a solitary and courageous pioneer who fought voluntarily to identify the headwaters of the Rio Verde and delimit the borders.

Guillobel reported the 1909 expedition in a document held in the Itamaraty archives (Guillobel 1910). He appointed a team, led by Captain Frederico Ferreira de Oliveira, to survey the Rio Verde and produce maps that would 'confront' those created by the Bolivian commission. In a letter to Guillobel, Oliveira outlined his plans for exploring the Rio Verde, referring to an agreement with the head of the Bolivian commission, General Pando. According to this agreement, the Bolivian sub-commission was to open a trail along the left bank of the Guaporé River up to the source of the Rio Verde. This trail would serve both to determine the longitude—using a new telegraphic connection with the town of Mato Grosso—and to facilitate the return of the joint sub-commissions, who would ascend the Rio Verde together (Guillobel 1910).

The longitude of the town of Mato Grosso was a significant point of contention between the boundary commissions, as the reference point used for triangulation could shift the perceived location of the source of the Rio Verde. In a document found in the Itamaraty Archive, based on a map produced by the French engineer Verreseu, the Brazilian commission sought

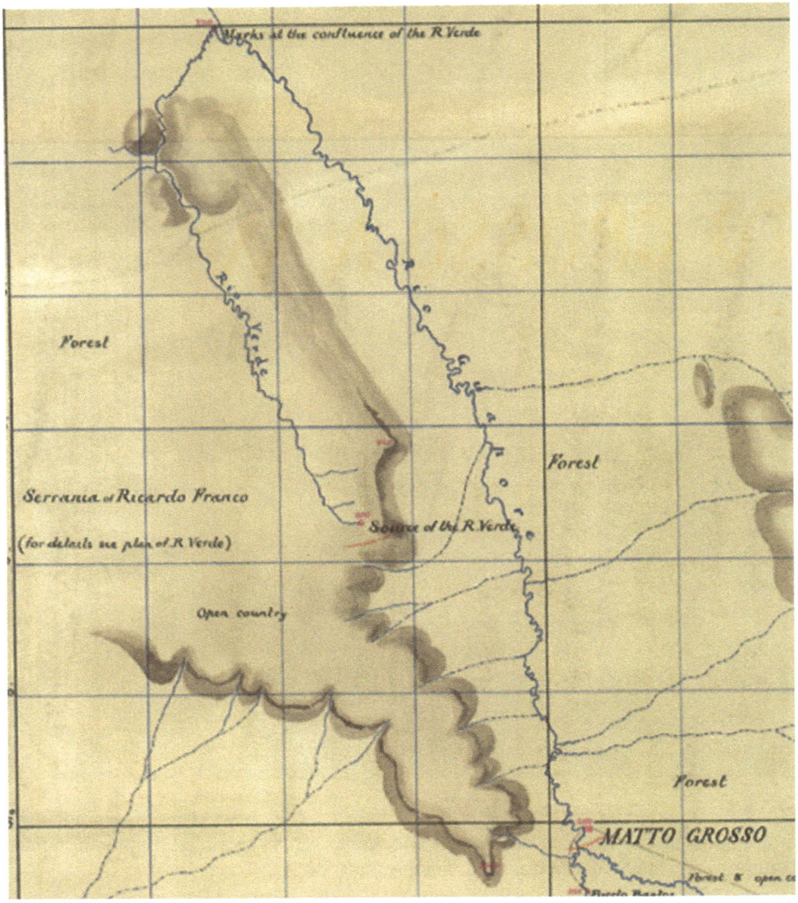

Fig. 7.7 Percy Harrison Fawcett, Matto Grosso, N. W. 1909. Detail from the River Verde. *Source* ARGS. ARGS [Collection of maps by Maj. P. H. Fawcett F.R.G.S, 1907–1910.] Printed with Permission

to challenge the position of the river's source (Fig. 7.8). The point marked 'A' represents the source of the Rio Verde as deduced from the 'erroneous longitude attributed by Major Fawcett to the old town of Mato Grosso'. The point marked 'B' indicates the source as deduced from the longitude of the Mato Grosso old town, this time determined by telegraphic means. By depicting the source of the Rio Verde controversy, the map suggests a greater accuracy of Brazilian maps, owing to the use of telegraphic methods in obtaining coordinates. Indeed, the coordinates for Mato Grosso differ between Guillobel's and Fawcett's records, and according to the annotations beneath the map, these discrepancies could account for a disputed area of 4334 km^2.

To provide an accurate location for the source of the Rio Verde, the commission convened in the town of Mato Grosso in June 1909, when Captain Frederico Ferreira de Oliveira reported Fawcett's arrival (Guillobel 1910). As agreed, the Bolivian commission was responsible for opening a path from the source of the river to the Guaporé. According to the report, Fawcett assured Oliveira that he would arrive at the Guaporé within 15–20 days, with the trail completed. Oliveira then set off with the Brazilian group, ascending the waterfalls along the Serra de Ricardo Franco. As had occurred with Fawcett, the conditions for mapping and exploration proved extremely challenging, and after several weeks navigating the waterfalls, Oliveira fell from a large rock. The incident was recorded both by the Brazilian commission and

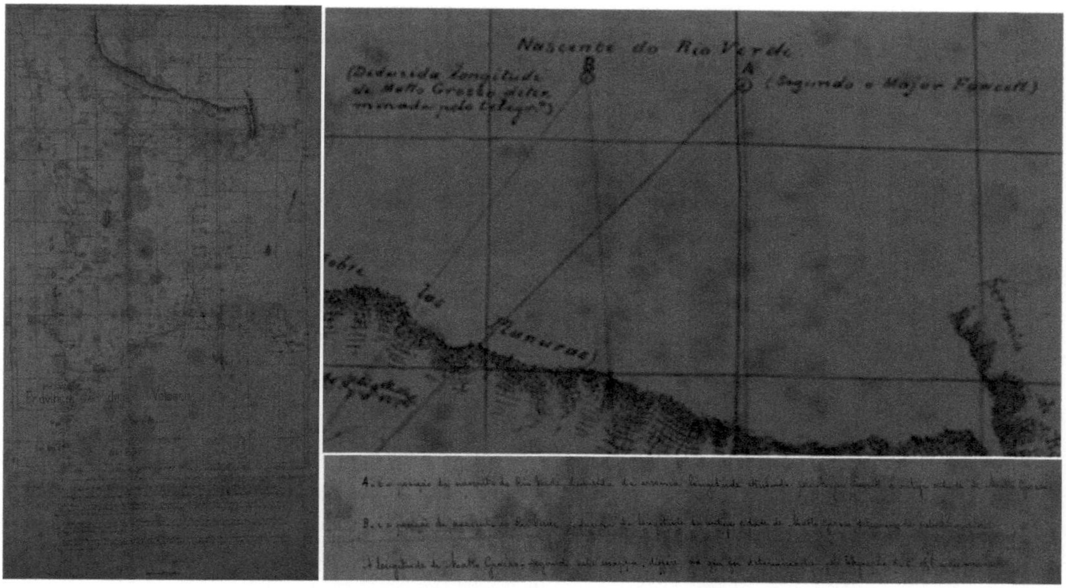

Fig. 7.8 Provincia de Velasco. Anonymous. The map contains a note regarding French engineer Verreseu, who made the map that served as a basis for the representation and some letters and captions to argue that Fawcett mistakenly placed the river sources in point 'A'. *Source* MHDI and MHI

by Fawcett, who noted that the Brazilian officer 'fell into the river, became feverish, and was forced to turn back', leaving the commission 'without instruments' (Fawcett 1910, p. 163).

Upon returning to Mato Grosso for medical attention, Oliveira was surprised to find that Fawcett had also returned to the town on 18 August, particularly as the English officer had failed to open the promised trail. According to Fawcett, his team had refused to carry out the work, fearing an Indigenous attack, and he had instead focused on positioning and constructing a boundary marker at the source of the Rio Verde (Guillobel 1910). In his report, Guillobel strongly criticised Fawcett's actions. According to the Brazilian officer, 'it was already known that Major Fawcett would not carry out his superior's orders' (Guillobel 1910), suggesting that Fawcett may have intended from the outset to abandon the trailblazing task to concentrate on surveying the terrain and legitimising his boundary marker. Following this contentious expedition, Fawcett produced his 1909 map of the Rio Verde headwaters (Fig. 7.9) and returned once again to England to present his findings at the Royal Geographical Society in 1910.

In a lecture delivered one year after his first expedition to the Rio Verde, Fawcett reinforced his role as the leading figure in mapping the region and made no mention of any maps produced by the Brazilian boundary commission. According to his account, 'the interior of South America remains largely unmapped, or incorrectly mapped, and in many areas, still unexplored' (Fawcett 1910, p. 515). However, despite this narrative, a closer examination of Fawcett's final map of the Rio Verde reveals a highly schematic and generic depiction of the terrain, particularly on the eastern face of the Serra de Ricardo Franco. This stylised representation stands in contrast to maps produced by the Brazilian commission and signed by Guillobel, in which the distinctive peaks of the hills are more clearly delineated and detailed than in Fawcett's versions (Fig. 7.10).

In another map attributed to the Joint Brazil–Bolivia Border Commission (Fig. 7.11), and also signed by Guillobel, Ricardo Franco's topography to the east of the Serra is represented with

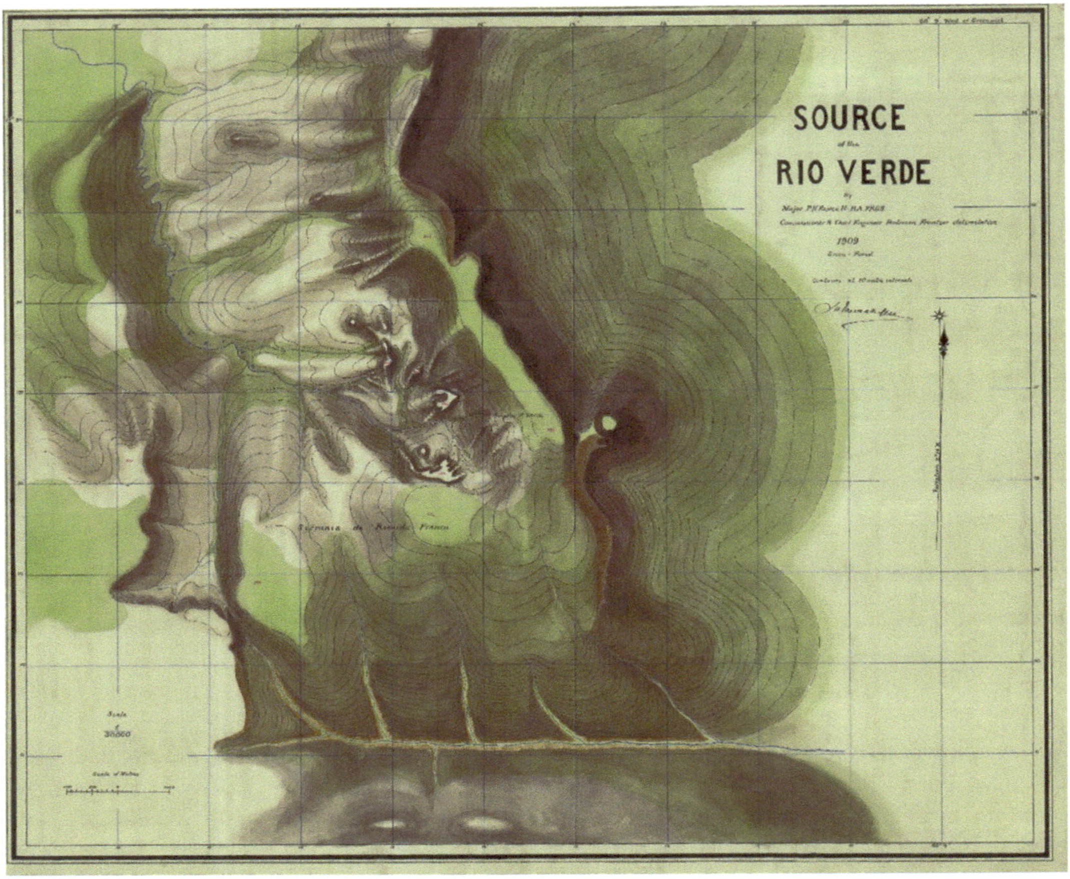

Fig. 7.9 Source of the Rio Verde, by Major P. H. Fawcett 1909. 1:30 000. 22 × 26 in. *Source* ARGS [Collection of maps by Maj. P. H. Fawcett F.R.G.S, 1907–1910.]. Printed with Permission

greater precision and detail. Just below the title, the map records the longitude and latitude of the Rio Verde's source, accompanied by an explanatory note stating that a marker had been erected at the site—graphically represented by a small triangle. This cartographic detail underscores the symbolic dimension of boundary-making through material inscriptions on the landscape. It is particularly revealing, however, to observe how the coordinates inscribed on Guillobel's map diverge from those registered in Colonel Percy Fawcett's field notebooks (Fig. 7.5). While the map bearing Guillobel's signature establishes the latitude of the source at 14° 57′ 15.2", Fawcett records a position of 14° 36′ 53"; and whereas the map notes a longitude of 16° 13′ 56.2", Fawcett gives 60° 16′ 29". According to the Brazilian commissioners, this discrepancy stemmed from Fawcett's triangulation method based on the city of Mato Grosso, a practice he explicitly describes in his own field notes. Such inconsistencies illuminate not only the technical challenges of geodetic observation in remote and densely forested areas but also the epistemic tensions inherent in international boundary surveys of the early twentieth century.

The comparison between the Brazilian and British records reveals more than a mere difference in measurements; it exposes the coexistence of distinct cartographic practices within the boundary commissions. It also challenges the colonialist narrative suggesting that Fawcett's maps filled a supposed void left by South American cartographers lacking the necessary exper-

Fig. 7.10 Limites do Brasil com a Bolívia. Rios Verde e Turvo. José Candido Guillobel, 1910. Signed by Guillobel on 31 March 1910. *Source* MHDI and MHI

tise or technical means. When Fawcett's 1909 map is compared with the commission's map signed by Guillobel (Fig. 7.12), it becomes evident that Fawcett produced a far more generalised representation. His use of regular curves conveyed only a schematic depiction of the relief, whereas Guillobel's map sought to present a more detailed and precise topographical rendering.

The representation of relief is visibly more detailed in the Brazilian maps than in those produced by the Bolivian commission under Fawcett. Yet it was Fawcett's maps that travelled to London, where they were adapted for publication in The Geographical Journal and archived at the Royal Geographical Society. The prestige of Fawcett's maps within the London-based epistemic community, however, did not mean that these images were widely accepted in the diplomatic negotiations over South American borders. The simplicity of Fawcett's maps led to them being received with scepticism by South American audiences.

Fawcett's controversies led his maps and landmarks to become a frequent target in Brazilian diplomatic reports. The reception of his maps could be approached by historical comments, such as the one published by the Brazilian General Sebastião Claudino de Oliveira Cruz in 1959. According to the General, the mistrust regarding Fawcett's mapping practices was established soon after the circulation of his documents. It is not by chance that the chancellors who requested the exploration of the Rio Verde (Rio Branco and Claudio Pinilla) met in 1911 and did not mention Fawcett's maps or the corrections of the source of the Turvo river. His maps were also ignored in the diplomatic presentation of Rio Branco to negotiate the new border in 1910.

According to Cruz, 'anyone with any geographical culture reading the documentation referring to the reconnaissance work on

Fig. 7.11 Levantamento das Cabeceiras do Rio Verde, 1910. Pela Comissão Mista de Demarcação de Limites. Signed by Guillobel. *Source* MHDI and MHI

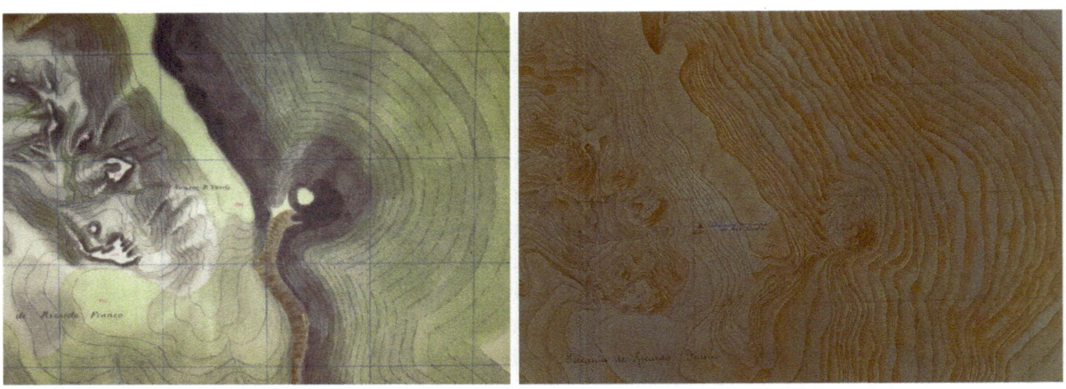

Fig. 7.12 Comparison of terrain representation in maps of the Rio Verde headwaters by Percy Harrison Fawcett (1909), on the left, and José Candido Guillobel (1910), on the right. Sources: Details from *Source of the Rio Verde* by Major P. H. Fawcett 1909, RGS, and from *Levantamento das Cabeceiras do Rio Verde*, 1910. Pela Comissão Mixa de Demarcação de Limites. Signed by Guillobel, MHDI and MHI

the River Verde in 1909 will immediately be dominated by the suspicion, almost certainty, of an improbable work carried out at that time' (Cruz 1959). Fawcett's map is, therefore, seen more as a rhetorical image for winning allies than the accurate and heroic work he celebrated in his accounts. Addressing the relationship between the Brazilian and the Bolivian commissions, Cruz argued that Fawcett carried out the 1908 exploration without communicating with the Brazilian group. According to Cruz (1959), 'in 1908, exclusively on behalf of the Bolivia commission, ahead of the reconnaissance operation in mixed form, the British explorer Major Percy Harrison Fawcett chose a source for the Verde River'. The world 'chose' was not placed by chance here, emphasising the mistrust in Fawcett's methods.

According to the Brazilian general, the commission from 1909 was also very problematic because, while a small Brazilian party accompanied Fawcett by land, the core of the Brazilian commission insisted on continuing by the river to follow the exploration protocols defined in a previous agreement. According to Cruz, Fawcett would have broken exploration protocols by going upriver by land, and the commissions could not meet at the source.

Despite all these controversies, Fawcett's map was an essential source for many Brazilian maps in the first decades of the twentieth century, and even the Brazilian Explorer Marechal Candido Rondon mapped the sources of the Verde following Fawcett's positions in 1935. A significant correction in the position of the source in Brazilian maps occurred only after 1946, when a Brazilian party corrected its coordinates and produced new maps. The mistake committed by Fawcett is recognised by his son, Brian, in a footnote of the book he arranged and published in 1953 from his father's materials. After many celebratory sentences about the accuracy of the border work in 1909, Brian Fawcett added:

> After all these sufferings here, my father would have been bitterly disappointed had he known that what he took to be the source of the Verde was not the true source at all. He followed what appeared to be the mainstream, but the Verde is a river of many small branches, and one of these opened out into a wide body of water not far from its mouth. The true source was discovered by Colonel Bandeira Coelho in 1946, some distance to the south-west of my father's position (Fawcett 2001 1953, p. 133).

After this exploration, according to Cruz, any Brazilian map portraying the source of the river Verde based on Fawcett's location will be 'making an unforgivable mistake'. The author quoted the map of Mato Grosso published in 1952 by Rondon, placing the sources with their new directions. Considering its pictorial function and guesswork, Fawcett's maps were progressively abandoned as references for border treaties. This process of disbelief in Fawcett's maps did not only occur in the Rio Verde, and the archival combination could be used as a method to challenge Fawcett's explorations and his pioneering mapping work on Amazonian rivers.

7.5 Conclusion

In concluding her article on the role of circulation in the production of geographical knowledge, Zusman (2015) draws attention to the fact that some ideas 'travel little, or not at all'. Depending on the agents and sites of production, certain forms of knowledge are more likely to be recognised as global, as they become incorporated into broader networks of territorial data collection and storage. This does not mean, however, that such knowledge is necessarily better, more accurate, or produced with greater methodological rigour. There are social conditions that shape the circulation of knowledge, meaning that some theories and territorial understandings are more likely to travel than others (Mignolo 2003a, b).

In the case of the maps produced by Fawcett and Guillobel, it becomes quite evident that the quality of representation does not determine the extent of their circulation. The fact that Fawcett was a British military officer meant that the maps of the Bolivian boundary commission were presented and republished in a major European centre of calculation, such as London, while the circulation of the Brazilian maps remained largely confined to South America. Fawcett's maps were adapted and published in journals, whereas Guillobel's remained unpublished, stored in the Itamaraty archives. This, however, does not mean that the maps endorsed and published by the Royal Geographical Society had a greater impact on the diplomatic negotiations concerning international boundaries, which were largely conducted based on Guillobel's maps.

Maps have never stood alone, their meanings are shaped by the images and texts we draw upon to interpret them. A map of the Rio Verde and its copy may be read as either an original scientific discovery or a questionable, biased representation, depending on the archive in which it is consulted. A transnational history of cartography, bringing into dialogue documents produced in different languages and from differing perspectives, can help to challenge the 'ideal of cartography' (Edney 2019), in which mapping is understood as a narrowly defined set of practices. Assumptions in the depiction of relief and errors in the triangulation of headwaters are not unique to nineteenth-century maps. As Fawcett claimed, such flaws could supposedly be 'corrected' by a heroic overseas explorer equipped with instruments. The use of instruments in difficult terrain, however, posed challenges for both the Bolivian and Brazilian commissions, demonstrating how scientific practices were subject to processes of calibration, adaptation and transformation.

As a participant in boundary commissions in the early twentieth century, Fawcett was undoubtedly a key figure in the mapping of South American frontiers. His contributions, however, like any other advance in the history of cartography, must be demonstrated, not presumed (Edney 1993). As I have sought to show in this text, placing archival sources in dialogue can be a powerful strategy for challenging a range of assumed narratives in the history of maps and in the production of geographical knowledge more broadly.

Archival Sources

ARGS Archives of the Royal Geographical Society, London.

ARSG Fawcett, Percy Harrison. The Fawcett Collection. 'The exploration of the river Aquiry, 1906-7'. Astronomical observations, Bolivia-Peru boundary and Bolivia-Brazil boundary, 1908-10. 'The new frontier between Bolivia and Peru', 1910. All RGS correspondence from Fawcett and his wife, and letters concerning the search for Fawcett. There are also extensive files of material, some 18 folders, about the search for Col. Fawcett. 1906-1988. Royal Geographical Society.

ARSG Fawcett, Percy Harrison. The new frontier between Bolivia and Peru Holo. 3 pp. foolscap. JMS/6/170 Section of the road of the Inca M.and R. Co. List of road distances from Tambopala to Astillero, 1 p. 1910.

MHDI Museu Histórico e Diplomático do Itamaraty (Historical and Diplomatic Museum of Itamaraty), Rio de Janeiro. MHI Mapoteca Histórica do Itamaraty (Itamaraty Historic Map Collection), Rio de Janeiro.

MHDI Guillobel, José Candido (1907). Limites setor Bolívia. Comissão de Limites Boliviana. Dificuldades surgindo dentro da comissão devido a oficiais ingleses. Museu Histórico do Itamaraty. Lata 460, Maço 25.

MHDI Guillobel, José Candido (1910). 1 comissário da Comissão de Limites do Brasil com a Bolívia. Limites do Brasil com a Bolívia. Relatório: Da exploração do Rio Verde da sua confluência com o Guaporé às suas cabeceiras e do terreno no qual, em 1877, foi levantado o marco das supostas nascentes daquele rio.

References

Craib R (2009) Relocating cartography. Postcolonial Stud 12(4):481–490

Cruz SCO (1959) Roboré, Infeliz Instrumento Diplomático. O Semanário, Rio de Janeiro, 5 a 11 de setembro de 175(IV):7–10

Edney MH (1993) Cartography without "progress": reinterpreting the nature and historical development of mapmaking. Cartographica 30(2–3)

Driver F (2001) Geography militant: cultures of exploration in an age of empire. Blackwell, London

Driver F, Jones L (2009) Hidden histories of exploration: researching geographical collections, RGS-IBG/Royal Holloway

Edney MH (2019) Cartography. The ideal and its history. University of Chicago Press, Chicago and London

Fawcett PH (1909) Survey work on the frontier between Bolivia and Brazil. Geogr J 33(2):181–185

Fawcett PH (1910) Explorations in Bolivia. Geogr J 35(5):513–529

Fawcett PH (2001/1953) Exploration fawcett. By Lt.-Col. P. H. Fawcett ... Arranged from his manuscripts, letters, log-books, and records by Brian Fawcett. Hutchinson, London

Fifer JV (1966) Bolivia's boundary with Brazil: a century of evolution. Geogr J 132(3):360–372

Ganzert FW (1934) The boundary controversy in the upper amazon between Brazil, Bolivia, and Peru, 1903–1909. Hispanic Am Histor Rev 14(4):427–449

Goes Filho SS (1999) Navegantes, bandeirantes, diplomatas: um ensaio sobre a formação das fronteiras do Brasil. São Paulo, Martins Fontes

Gole S (1989) Indian maps and plans from earliest times to the advent of European surveys. Manohar, Mahon Delhi

Gunther RT (1932) The astrolabes of the world. Oxford University Press, Oxford, vol I, pp 179–228

Hägerstrand T (1952) The propagation of innovation waves. Lund studies in geography: Series B, Human geography, 4. Royal University of Lund, Department of Geography, Lund

Hodder J, Heffernan M, Legg S (2021) The archival geographies of twentieth-century internationalism: nation, empire and race. J Hist Geogr 71:1–11. https://doi.org/10.1016/j.jhg.2020.06.008

Jöns H, Heffernan M, Meusburger P (2017) Mobilities of knowledge: an introduction. Springer, London

Kain RJP, Baigent E (1992) The cadastral map in the service of the state: a history of property mapping. Chicago & London: University of Chicago Press

Konishi S, Nugent M, Shellam T (2015) Exploration archives and indigenous histories: an Introduction. In: Konishi S et al (eds) Indigenous intermediaries: new perspectives on exploration archives. Australian National University Press and Aboriginal History Inc., Sidney, pp 1–11

Machado LO (2000) As idéias no lugar, Terra Brasilis [Online] 2. https://doi.org/10.4000/terrabrasilis.298

Massey D (2008) Pelo espaço. Bertrand, Rio de Janeiro

Mignolo W (2003a) Historias locales/diseños globales. Colonialidad, conocimientos subalternos y pensamiento fronterizo. Akal, Madrid

Mignolo W (2003b) La colonialidad a lo largo y a lo ancho: el hemisferio occidental en el horizonte de la colonialidad moderna. In: Lander E (ed) La colonialidad del saber: eurocentrismo y ciencias sociales. Clacso, Buenos Aires, pp 55–85

Moraes ACR (1990). Ratzel. Ed. Atica, São Paulo

Novaes AR (2020) Mapping cross-cultural exchange: Jaime Cortesão's dialogues and documents on the role of indigenous knowledge in Brazilian exploration In: Schelhaas B et al (eds) Decolonising and Internationalising geography: essays in the history of contested science. Springer, Cham. pp 1–16

Novaes AR (2020). Histórias Escondidas nos Mapeamentos Fronteiriços. Terra Brasilis 14. https://doi.org/10.4000/terrabrasilis.7124

Phillimore RH (1952) Three Indian maps. Imago Mundi 9:111–114

Pratt ML (1993) Criticism in the contact zone: decentering community and nation. In: Bell S et al (eds) Critical theory, cultural politics and Latin American narratives. University of Notre Dame Press, Notre Dame, pp 83–102

Quijano A (2000) Colonialidad del Poder, Eurocentrismo y América Latina. In: Lander E. (ed) La Colonialidad del Saber: Eurocentrismo y ciencias sociales. Perspectivas latinoamericanas. Clacso, Buenos Aires

Raj K (2007) Relocating modern science: circulation and the construction of scientific knowledge in South Asia and Europe, 1650–1900. Palgrave Macmillan, Basingstoke

Ratzel F (1876) Die chinesische Auswanderung. Ein Beitrag zur Cultur- und Handelsgeographie. Kern, Breslau

Resor RR (1977) Rubber in Brazil: dominance and collapse, 1876–1945. Bus Hist Rev 51(3):341–366

Said EW (1983) The world, the text, and the critic. Harvard University Press, Cambridge

Said EW (2004) Teoría Ambulante. In: El mundo, el texto y el crítico. Debate, Madrid, pp 303–330

Sauer C (1952) The agricultural origins and dispersal. The American Geographical Society, New York

Seemann J (2012) Friedrich Ratzel entre Tradições e Traduções. Terra Brasilis 1. https://doi.org/10.4000/terrabrasilis.180

Subrahmanyam S (1997) Connected histories: notes towards a reconfiguration of early modern Eurasia. Mod Asian Stud 31(3):735–762

Vallaux C, Brunhes J (1921) La Geographie de L'Histoire. Geographie de la Paix et de la Guerre Sur Terre et Sur Mer. Libraire Félix Alcan, Paris

Vergara M de R (2010) Ciência, fronteiras e nação: comissões brasileiras na demarcação dos limites territoriais entre Brasil e Bolívia, 1895–1901. Bol Mus Para Emílio Goeldi 5(2):345–361

Wimmer A, Glick Schinder N (2002) Methodological nationalism and beyond: national state building, migration and social sciences. Glob Netw 2(4):301–334

Zusman P (2015) No solo el reposo configura geografías. Terra Brasilis 5. https://doi.org/10.4000/terrabrasilis.1367

Divide and Rule: Border and Identity in Late Nineteenth-Century Portuguese Cartography

8

Luís Miguel Moreira

Abstract

In 1864, the need to define the precise boundary between Portugal and Spain—the Raia—prompted the production of specific topographical maps, under the supervision of the Corpo do Estado-Maior (General Staff Corps). The main objective of this initiative was to define the jurisdictions of the two states and put an end to legal ambiguities and conflicts. This process required scientific cartographic mapping, and to achieve this, the Comissão Internacional de Limites (International Boundary Commission) produced and published three general map series of the border, at different scales. For the first time, the frontier was clearly defined, demarcated and systematically documented. The main aim of this analysis is to compare the way the border was depicted in these first official maps, which were produced for political and diplomatic negotiations, with other more widely circulated maps, such as military and civilian topographic series. It considers how these representations adopted and/or conveyed geopolitical narratives, with regard to the quantity, quality and consistency of information included; identifying evidence of conflict, state presence, or local cooperation and determining the usefulness of the data presented and the intended users. Once the boundary line was drawn, central government authority extended along the border.

8.1 Introduction

From the sixteenth century until the end of the eighteenth, the main European kingdoms initiated a process of centralising power, replacing the earlier feudal political structure. This shift necessitated a redefinition of the concept of the 'border', and the criteria by which borders were to be delineated (Conrad and Křížová 2023). As such, natural frontiers and military boundaries once used to define a monarch's jurisdiction were gradually supplanted by new legal and diplomatic concepts aimed at establishing the territorial limits of state sovereignty (Stoian 2014).

In this way, borders (or border areas), which had previously been more fluid and flexible, became more stable and clearly defined from the eighteenth century onwards—not least because, from the perspective of the central state, they

This work has been supported by the Spanish State Research Agency under Grant Number PID2020-114088GB-I00/AEI/10.13039/501100011033

L. M. Moreira (✉)
Geography Department, Institute of Social Sciences, University of Minho, Guimarães, Portugal
e-mail: lmmgeo@gmail.com

Centre of Geographical Studies, Institute of Geography and Spatial Planning, University of Lisbon, Lisbon, Portugal

J. García-Álvarez and A. Reyes Novaes (eds.), *Mapping the Contested Line*, Historical Geography and Geosciences, https://doi.org/10.1007/978-3-032-29889-8_8

began to play a more active role, particularly in the collection of taxes and customs control.

Following the Napoleonic Wars in the nineteenth century there was a significant redrawing of borders, as new territorial entities emerged and disappeared. In Central and Eastern Europe especially, imperial powers consolidated their domains. Alongside this, a nationalist political movement, shaped by Romantic ideals and issuing in part from the consolidation of the apparatus of the state, swept across the European continent.[1] Communities began asserting their national identities based not only on language, history, culture and religion, but above all on the identification of a common territory—a homeland—delimited by clearly identified, stable borders. Defining the national space thus came to play a fundamental role in the construction of national identity (Biggs 1999; Halmos et al. 2023).

By the close of the century, the concept of the nation-state, in which the political borders of the state were one and the same as the cultural boundaries of the nation, had become the preferred model of sociopolitical organisation across Europe (Hagen and Diener 2012).

In this wider context, it became a priority to identify, describe and record territory through systematic, scientific cartography, proceeding from this foundation to demarcate its borders (Goettlich 2019). The more recent advances in geodesy and topography provided the technical means to standardise methods for mapping and defining borders between centralised states—an essential condition for the construction of the nation-state.

At the start of the nineteenth century, France was the only European country to possess a general topographic map (at a scale of 1:86,400). It was based on astronomical measurements and a geodetic network formed by a chain of triangles, and had been produced by four generations of the Cassini family, starting in the second half of the seventeenth century (Pelletier 2019).

Although it had been a private initiative, both commercially and academically motivated, the map demonstrated clear military utility, as well as addressing the needs of the public administration, notably as a tool to support development policies and territorial reform. As such, following the French Revolution of 1789, it was nationalised and became an official state instrument (Konvitz 1987).

By representing French territory according to previously defined scientific criteria (astronomical, geodetic and topographic) and standardising both cartographic language (symbols, relief depiction, typography, etc.) and language use (employing only standard French), the Cassini Map helped to construct a collective mental map of France and promote a national territorial identity (Brotton 2019).

Despite the many errors identified in this map, which eventually led to the French authorities commissioning a new topographic survey, over the first decades of the nineteenth century this project became the model for European cartography. Almost always directed by military institutions, topographic mapping became a state responsibility. By the middle of the century, almost all the European powers had already begun, or in some cases completed, the process of surveying for a general topographic map capable of meeting the primary political and administrative needs of government, including territorial division and the definition of international borders.

Generally, in Spain, the UK and Portugal (as we shall see), the approach for topographic map production, although state-led—under the responsibility of Spain's Instituto Geográfico Nacional (National Geographical Institute) from 1870, or Britain's Ordnance Survey from 1790—followed a different cartographic model from the French and other European states. It was civilian institutions that were responsible for both the topographical survey and the land registry that accompanied it, even though the technical staff was almost entirely made up of military officers (Nadal and Urteaga 1990).

For the Portuguese authorities, defining the country's land borders was a prerequisite for

[1] The role of the state in constructing the idea of the nation is embodied in what Hobsbawm and Ranger (1988) define as the 'invention of tradition'. More recently, this theme has been revisited López Facal (2013).

drawing up an official map. From the sixteenth century to the end of the nineteenth, the Portuguese state sought more rigorous and precise knowledge of the country to improve its administration's efficiency and to legitimise its sovereignty within Iberia. Until the mid-nineteenth century, however, the production of maps of the whole of Portugal was essentially a business dominated by foreign geographers, cartographers and publishers. The Portuguese commonly used hand-drawn copies and versions of foreign printed copies, corrected and updated to varying degrees, which were used as official cartography by the political, diplomatic and military authorities.

Thus, the independent production of national maps evolved alongside the establishment of political borders. To respond to both these developments, Portuguese cartography needed to represent the territory accurately, using topographic, chorographic, geodetic and astronomical techniques (Branco 2003). Only a scientific determination of boundaries was seen as capable of providing a uniform image of the entire territory of the country and avoiding recourse to less rigorous sources.

From the second half of the eighteenth century onwards, maps played a critical role in shaping geographical imagination and creating national identity across Europe. By standardising the representation of physical and human geography, maps helped bind populations to their natural environments—that is, their homeland—and sowed the seed for future nationalist movements (Wintle 2016).[2] After the signing of the Treaty of Vienna in 1815, the linear national borders established by new states served, above all, to differentiate ethnolinguistic and cultural communities, and to protect and maintain their distinct sociopolitical systems. An example is the border between France and the UK of the Netherlands, ratified by the Treaty of Courtrai in 1820, which covered the territory of Belgium and Luxembourg after four years of topographic surveys and diplomatic negotiation (Vandenbroucke and D'heygere 2021).[3]

Compared to the rest of Europe, the border between the kingdoms of Portugal and Spain was older and more stable. Disputed areas were relatively few, sparsely populated and usually the subject of local contention over property rights, land use, or smuggling. In this sense, the linear demarcation of the Luso-Spanish Raia (the name given to the border between the two countries, meaning 'the stripe') responded to the need to assert sovereignty, reflecting yet another stage in the process of establishing the apparatus of the liberal state (García-Álvarez and Santos Sánchez 2023).

In the Portuguese context specifically, 'the symbolic function of the border was one of shaping an identity-based nationalism steeped, in the case of the Portuguese, in pronounced anti-Spanish sentiment, with the imaginary dividing line coming to represent real separation' (Cairo et al. 2009).[4]

The cartographic definition of the border thus assumed particular importance, establishing the country's contours and separating it from the rest of Iberia, reinforcing the official geopolitical discourse. At a time when various 'Iberianist' currents were calling for Iberian integration—whether through dynastic union (monarchical), or federal structures (republican)—maps served to illustrate that Spain's political borders did not correspond to the geographic limits of the Iberian Peninsula, and that Portugal was not a Spanish province.

This article is structured in two parts. The first offers a general overview of the development of

[2] On the contribution of cartographic representation and geographical knowledge to national identity, see the case of Scotland, as analysed by Withers (2001). For the case of Spain, see García-Álvarez (2013).

[3] It is also worth noting that, in the second half of the eighteenth century, both Spain and France also undertook border demarcations, with diplomatic negotiations relying heavily on maps. Examples include the Convenio de Límites of Coll de Pertús, signed in Perpignan in 1764, which addressed the border in a small sector of the eastern Pyrenees, and the Treaty of Aranjuez of 1777, which defined the borders between the two kingdoms on the island of Santo Domingo. See García Álvarez and Puente Lozano (2023).

[4] Our translation. Unless otherwise stated, all translations of cited foreign language material in this chapter are our own.

Portuguese cartography in the second half of the nineteenth century, focusing in particular on the origins of three map series produced in Portugal by both civil and military institutions.

The second part outlines the key features of each of these maps and, finally, provides a comparative analysis of the border as it appears in the selected cartographic examples. In this way, the maps—at the time considered scientific documents presumed to offer neutral representations of geographic space—are examined side by side to uncover both explicit and implicit geopolitical narratives, conveyed through the 'silences' of cartography, as first theorised by Harley (1988a, b).

8.2 Portuguese Cartography in the Second Half of the Nineteenth Century

8.2.1 The Representation of Borders in Portuguese Cartography Before the 1864 Spanish-Portuguese Boundary Treaty

In maps of Portugal produced from the mid-sixteenth century to the end of the nineteenth, the border was represented using a wide variety of line types—solid, dashed, dotted, or coloured—and seldom with any consistency. As a result, it was virtually impossible to identify a definitive line (Garcia, 1996).

In this context, the role played by the 1864 Treaty of Lisbon (or Treaty on Boundaries) between Portugal and Spain is particularly important. It aimed to scientifically delimit and demarcate a single border line separating the two Iberian kingdoms, and ultimately made it possible to stabilise and standardise the cartographic representation of the country's external boundaries.

It is worth noting that, in the development of Portuguese cartography, there was already a tradition of representing the Raia on maps, particularly from the second half of the seventeenth century, in the context of the Portuguese Restoration War (1640–1668). Since most of the military activity took place in this part of the Iberian Peninsula, there was an increase in the production of maps, especially hand-drawn ones, which were used to support military decision-making. This meant that representation of the border served an exclusively military purpose (Alegria et al. 2007).

From the second half of the eighteenth century onwards, Portugal and Spain gained extensive practical experience in delimiting linear borders across their vast colonial territories in South America. In these negotiations, it became common practice to use maps as diplomatic documents for defining political boundaries, which were formalised in the treaties of Madrid (1750), San Ildefonso (1777) and El Pardo (1778) (Magalhaães et al. 1997). Maps also proved to be valuable tools for the various *partidas* (the forerunners of future international joint commissions) tasked with demarcating the boundary lines agreed upon in ministerial offices (Ferreira 2001).

This combination of scientific theory and practical experience would go on to serve as a model for application in the Iberian Peninsula (Branch 2010). Indeed, during the first decade of the nineteenth century, the governments of the two Iberian kingdoms felt the need to resolve certain territorial disputes along sections of their shared border. As a result, between 1803 and 1806, two international boundary commissions were established (composed of diplomats and cartographers), with the aim of adjusting the border line in two contested areas. The first was created along the Contenda de Moura (the Alentejo-Andalusia border), while the second was set up along the disputed area of Montes da Madalena-Lindoso (the Minho-Galicia border) (Dias 2009).

This process—based on the on-the-ground cartographic surveys, the agreement of the boundary line by the two commissioners, the production of technical reports supported by jointly prepared maps, and the subsequent submission of these materials to the respective ministries for the final political and diplomatic negotiations—laid the groundwork for the commissions which, from the second half of the century onwards, would establish the modern border between Portugal and Spain.

These initiatives—along with the project to reform the administrative division of *comarcas* (regional judicial-administrative districts) (Silva 1998) which was likewise based on the production of maps—formed part of the set of state functions characteristic of a 'cartographic state', whose objective was to know and construct the territory (Catroga 2013). Indeed, in the final decades of the eighteenth century, the Portuguese central administration had already recognised the value of maps as instruments of power. As a result, there was a growing desire, both among government officials and within the Portuguese scientific community, to produce a topographical map of the kingdom that could serve as a tool for territorial administration, economic development and military organisation, drawing on the latest advances in geodetic science and modelled after the British Ordnance Survey.

Thus, from 1790 onwards, the government, under the initiative of Luís Pinto de Sousa Coutinho (1735–1804), Minister and Secretary of State, resolved to take charge of this undertaking. The astronomer Francisco António Ciera (1763–1814) was appointed as its coordinator, and the other members of this 'geodetic commission' were the military engineers Carlos Frederico Bernardo de Caula (1766–1835) and Pedro Folque (1744–1848).

Geodetic observations began immediately and continued almost uninterrupted until 1803. The commission selected the points for fixing the vertices of the triangulation and measured the sides of the principal geodetic triangles, until the work was halted in the early years of the nineteenth century. An initial campaign between 1790 and 1791 enabled a survey of the territory south of the River Tagus, and during a foray into Galicia, in collaboration with some Spanish officers, it was possible to lay the groundwork for a future connection between the two Iberian geodetic networks (Dias 2003). At the same time, despite various obstacles, some topographic survey work was carried out in the Lisbon region, resulting in the production of around twenty sheets at a scale of 1:10,000—a good sample of what the intended final result may have looked like.

The sociopolitical and military upheavals that followed these two early attempts—triggered by the Napoleonic Wars (1795–1814), as well as the civil wars that ensued after the liberal revolutions in Spain (1812) and in Portugal (1820)—prevented the full realisation of this initial effort.

It was only in the second half of the nineteenth century, during a period of political calm, that the two Iberian kingdoms decided to agree on the definition of their shared land borders, as a consequence of the process of constructing centralist liberal regimes (Pro Ruiz 2007).

Once the principle had been established that sovereignty resided in the 'nation' and was exercised by its elected representatives, it became necessary to reform the political and administrative geography of each country, both internally and externally. This reform demanded the acquisition of precise knowledge of the territory that would serve as the foundation of the nation, with particular emphasis on cartographic representations: 'The dynamics of centralisation and modernisation of the state apparatus required geographical and statistical information, both about the territory (its configuration, orography, administrative boundaries, border line, agricultural production), and about the people and their activities—in short, about everything that fell under its remit. As previously noted, to fill these gaps, the state developed 'geographical information policies'. (…) This was the case with the *Carta geral do Reino* (General Map of the Kingdom)' (Branco 2003).

Internally, the reform of political and administrative division aimed to make both central and local administration more efficient, to ensure a more rapid application of justice, and to promote the creation of electoral constituencies adapted to the needs (and interests) of the new liberal regime. After various attempts and proposals, a territorial organisation model was finally approved in late 1836—one that would become established over the following decades and endure to the present day.

So, by the mid-nineteenth century, the nation was already organised internally, with political representatives linked to specific territorial units. However, its external boundaries had yet to be rigorously defined. It was still necessary to resolve sovereignty 'grey areas' and other borderland issues to unequivocally establish the national territory and thereby complete the process of state-building and constructing territorial identity.

To define these boundaries, geographical knowledge of the territory—often expressed in cartographic language with topographical features—was essential. The problem, however, was that the country was lagging behind in its process of geodetic and cartographic modernisation when compared to the efforts undertaken by other European states. In the mid-nineteenth century, General Augusto Xavier Palmeirim (1808–1890) summarised the situation in the following terms: '[…] yet no plan has so far been established among us, nor any systematically conducted work undertaken, that might have led us to a full understanding of the country; and the military remain deprived of a good map on which they might agree or design any war plan, or assess and become familiar with whichever map might best reflect the contours of the Portuguese landscape'(Palmeirim 1845).

At last, the moment had come for the Portuguese state to embark on the reform of national cartography—without which it would not be possible to establish, on scientific grounds, a common political border line between the two states, nor to present citizens (and not only the elites) with an unambiguous image of their homeland: one that could be recognised by all and that was capable of forging an identity-bearing and binding connection between the territory and the people who inhabit it.

On the contribution of maps to the formation of Portuguese national identity, it is worth noting that besides being used as technical instruments in the process of demarcating borders between states, maps were taken on a broader dimension almost as an ideological instrument and as a symbol alongside the flag and the anthem. '[…] Maps especially those that allow for the visualisation of the national territory in its entirety mark the boundary between 'us' and 'them', reifying a line of separation which, although it must be valued from a technical point of view, more importantly signals the limits of political belonging' (Cunha 2007).

8.2.2 Civil Cartography

The institutionalisation of scientific (terrestrial) cartography in Portugal took place over the course of the nineteenth century. In the initial phase, cartographic surveying remained within the military sphere, with responsibility first divided between the Corpo de Engenheiros Militares (Corps of Military Engineers, established in 1812) and, later, the Corpo do Estado-Maior (General Staff Corps, created in 1834).

At the same time, the country's development—based on the construction of transport infrastructure and the implementation of numerous public works projects, including the improvement of seaports, mining operations, agroforestry development and so forth—required a more precise and accurate cartographic representation of the territory, at various scales, supported by a multi-order geodetic network.

A large number of civil cartographic surveys were launched at this time (although, at first, the technical personnel responsible were military), formally coordinated by the Comissão dos Trabalhos Geodésicos, Topográficos e Cadastrais do Reino (Commission for Geodetic, Topographic and Cadastral Works of the Kingdom), which was incorporated into the Ministério das Obras Públicas, Comércio e Indústria (Ministry of Public Works, Commerce and Industry) in 1852. Given the financial, technical, human and material constraints, the final result took the form of a cartographic series—the *Carta geral do Reino*, or *Carta chorographica de Portugal* (Chorographic Map of Portugal), at a scale of 1:100,000 (although field surveys were conducted at a scale of 1:50,000)—whose editorial process extended from 1853 to 1904 (Branco 2003).

This ultimately became the most significant cartographic achievement of nineteenth-century Portugal, earning international recognition at the International Geographical Congress held in Paris in 1875. It was an atlas consisting of 37 pages, each measuring 50 by 80 cm, using the Bonne projection, the Puissant ellipsoid and the Lisbon datum. The ability to engrave and print (or lithograph) the sheets as they were completed marked a major paradigm shift in Portuguese cartography: whereas previously maps were still drawn by hand and treated as 'state secrets' with limited circulation, they now became public and accessible to a wide range of users—particularly within civil society.

Particular attention was paid to the quality of the engraving and printing of the sheets (which were manually lithographed), and each one recorded the principal features of the country's physical and human geography—including the hydrographic network, altitude (represented for the first time through contour lines), landforms and vegetation, as well as settlement centres, transport routes and toponymy. This series also served as the basis for various forms of thematic cartography, such as the *Carta agrícola e florestal de Portugal* (Agricultural and Forestry Map of Portugal), directed by Gerardo Pery and the *Carta geológica de Portugal* (Geological Map of Portugal), coordinated by Carlos Ribeiro and Joaquim Filipe Nery Delgado, published in 1876.

From that point onwards, national cartographic production was divided into two branches: the civil branch, located within the Direcção-Geral dos Trabalhos Geodésicos, Corográficos e Hidrográficos do Reino (Directorate-General for Geodetic, Chorographic and Hydrographic Works of the Kingdom) and its successors; and the military branch, initially under the responsibility of the Corpo do Estado-Maior and later of the Serviços Cartográficos do Exército (Army Cartographic Services).

8.2.3 Military Cartography

In the final quarter of the nineteenth century, the Corpo do Estado-Maior, recognising that the army lacked both small-scale maps for depicting the country's main overland routes and more detailed maps of strategically important areas, and noting that the *Carta chorographica* was slow to be completed and that its 1:100,000 scale offered limited topographic detail, sought to fill these gaps by supporting the development of two exclusively military cartographic projects: the *Carta itinerária* (Itinerary Map), at a scale of 1:250,000, and the *Carta dos arredores de Lisboa* (Map of the Environs of Lisbon), at a scale of 1:20,000.

The latter was first published in 1891, but it was only from the late 1920s onwards that surveys were extended to the entire country, under the title *Carta militar de Portugal* (Military Map of Portugal), at a scale of 1:25,000—a series that today still constitutes Portugal's principal base map (Dias and Botelho, 1999).

It was decided that the *Carta itinerária* should correspond to the boundaries of the military divisions, drawing on military reconnaissance missions and the survey work carried out for the *Carta chorographica.* Its publication began in 1881 with the sheet corresponding to the 1st Military Division (districts of Lisbon, Leiria and Santarém); this was followed by the 2nd Division (districts of Coimbra, Aveiro, Viseu, Guarda and Castelo Branco) the 4th Division (districts of Portalegre, Évora, Beja and Faro), between 1883 and 1896 and finally the 3rd Division (districts of Bragança, Vila Real, Porto, Braga and Viana do Castelo), in 1899–1900 (Dias 2018).

This first edition was followed by another, published between 1898 and 1900, consisting of nine sheets arranged according to a scheme that divided the mainland national territory into rectangular sections. This was soon followed by further versions: one with 12 sheets (between 1903 and 1921), and another with 29 sheets (post-1920), each with multiple editions. These would later serve as inspiration for the production of the country's first road maps.

The maps in this series were printed in colour, making them visually more appealing. In addition to information on road and railway routes connecting the main towns and cities, they also included a wide range of other features such as altimetry (with a contour interval of 100 m), orography, hydrography, vegetation cover and bathymetry. As the border demarcation work was already well advanced—and given that this was an area of military interest—the representation of the frontier was complete.

Meanwhile, in 1932 as the newly established Serviços Cartográficos do Exército (Army's Cartographical Services) began publishing the *Carta militar* at a scale of 1:25,000, it was decided to make use of all the collected information to update the *Carta itinerária.* In practice, rather than a simple update, a new cartographic series—M586—was proposed, comprising a smaller number of sheets (eight), under the title *Carta militar de Portugal*, using the Gauss projection system, while retaining the 1:250,000 scale. Production began in 1947, but only a few sheets were completed at the time; for various reasons, the project was repeatedly delayed and only fully completed between 1961 and 1968.

8.2.4 Border Cartography

The entire process of modernising civil and military cartography coincided with the definition of the land border. Initiated in the late 1860s, the delineation of the Luso-Spanish Raia prompted the production of specific cartography with topographic characteristics, also under the responsibility of the Corpo do Estado-Maior.

While the preparatory work for the signing of the Treaty of Lisbon (or Treaty on Boundaries, of 1864) lasted around a decade and involved inventorying the available (historical) cartographic sources and the surveying and mapping of certain sections of the border where sovereignty was in doubt—particularly along the northern frontier with Galicia—the subsequent demarcation of the dividing line continued for the next few decades and unfolded in two main phases, marked by significant interruptions. The first phase lasted until 1906, the year the *Acta geral da delimitação entre Portugal e Espanha* (General Act of Delimitation between Portugal and Spain) was signed in Lisbon, establishing the boundary between the mouth of the River Minho and the confluence of the River Caia with the River Guadiana. The second phase, twenty years later, culminated in the signing of the *Convénio de limites entre Portugal e Espanha* (Convention of Limits between Portugal and Spain), signed in Lisbon in 1926, which fixed the border line between the confluence of the River Cuncos with the River Guadiana and the mouth of that river. In the intervening period, most of the major border disputes were resolved, including the contested areas of Couto Misto in the north and the Contenda de Moura in the south—the latter having been settled by the convention of 27 March 1893 and the demarcation act of 18 July 1894 (Santos-Sánchez 2022).

The major exception—which remains unresolved today—is the territory corresponding to the former municipality of Olivença, whose border delimitation continues to lack formal agreement between the two countries (Pinto 2022).

This laborious and lengthy process required detailed scientific cartographic description at various scales (Garcia Álvarez and Puente Lozano 2015). In this regard, the International Border Commission between Portugal and Spain, in addition to its extensive manuscript output, published three general cartographic series of the demarcated border up to 1906: the *Plano da linha de fronteira entre Espanha e Portugal* (Plan of the Border Line between Spain and Portugal), at a scale of 1:10,000, published between 1896 and 1905 to accompany eight partial border handover proceeding, and which is divided into eight sections and comprises 194 sheets, each covering a strip of land 400 m wide (200 m on either side of the line); the *Carta chorographica dos limites de fronteira* (Chorographic Map of the Border Limits), at a scale of 1:50,000, composed of 22 sheets, representing a strip of land eight kilometres wide (four kilometres on each side of the line). A first complete version of this series was published in 1893, then revised and reissued in 1904; and the *Planta geral da linha de fronteira desde a foz do rio Minho até à confluência do rio Caia no Guadiana* (General Map of the Border Line from the Mouth of the River Minho to the Confluence of the River Caia with the Guadiana), at a scale of 1:100,000, consisting of 11 sheets and annexed to the *Acta geral da delimitação entre Portugal e Espanha* of 1906.

Additionally, the commission prepared the *Planta do Rio Minho desde sua foz até a confluência do rio Troncoso* (Map of the River Minho from its Mouth to the Confluence with the River Troncoso), at a scale of 1:2500, consisting of 59 sheets and published in 1898. This served to complement the handover record of the border line along the River Minho.

Much of this surveying work was based on the Portuguese geodetic network, initially established in the 1790s but revised and completed from the 1840s onwards. Its first- and second-order triangulation supported the ongoing chorographic surveys, which were being carried out simultaneously, specifically at a scale of 1:50,000. Given the topographic features of the region, the maps already showed a considerable level of detail in the depiction of the terrain, despite the limited extent of the mapped area. Unsurprisingly, the depiction of the border line became the principal focus of the cartographic technicians.

Finally, after nearly half a century of work, the border was scientifically described, delimited and demarcated (Garcia 2000).

8.3 Cartographic Representations of the Borderland–An Interpretation (Part One: 1864–1926)

Drawing on the catalogued Portuguese cartographic sources—both the printed 'official' maps produced for diplomatic use and the maps in general circulation, of civil or military origin—this section aims to interpret representations of the border and the organisation of the Raia area. It seeks to address several questions concerning the quantity, quality and consistency of the information represented on either side of the border line; identifying evidence of conflict, state presence, or local cooperation and determining the usefulness of the depicted information and its intended audience (users).

The civil cartographic source used is the *Carta chorographica de Portugal*, at a scale of 1:100,000 (also known as the *Carta geral do reino*). Its 37 sheets bear different survey and publication dates, as the work extended over several decades (1853–1904), and it served as the basis for the surveying and production of other cartographic series, such as the *Nova carta chorographica de Portugal* (New Chorographic Map of Portugal), at a scale of 1:50,000, produced at the beginning of the twentieth century.

The military source used is the *Carta itinerária*, produced at a scale of 1:250,000, in the version composed of 12 sheets, published between 1903 and 1921, and made accessible to the public through its automotive edition. Finally, the *Carta chorographica dos limites de fronteira*, published in 1893 and revised from 1906 onwards, at a scale of 1:50,000, constitutes the political, diplomatic and military version of the border line—by then already defined and fully demarcated. It serves as the official record of the border line.

Given the length of the Raia and the large number of cartographic documents consulted, it was decided to delimit a number of sample areas along the Luso-Spanish border that would, on the one hand, be sufficiently representative of the whole—including both the Raia Húmida ('Wet Raia' following the course of the River Minho) and the Raia Seca ('Dry Raia') on the part of the border between Portugal and the Spanish province of Galicia—and, on the other hand, focus on areas that, over the course of history, have been the subject of disputes, friction, or some form of conflict between the two Iberian countries.

8.3.1 The *Carta Chorographica De Portugal*, 1:100,000

The first impression that emerges from an analysis of the 19 sheets in this cartographic series which depict both the border line and the Raia—understood here as a strip of territory on both sides of the border, of varying width—is that its representation is neither consistent nor uniform.

The main reason for this imbalance lies in the slowness of the surveys, some of which were initiated even before the first border demarcation work began, and in the fact that both projects—the *Carta chorographica* and the border demarcation—developed simultaneously.

This means that the border is marked with a dashed line on only five of the sheets, and on one of them, the line demarcates only the Raia Seca. Most sheets depict the terrain on the Portuguese side, leaving the corresponding Spanish territory blank, although some include a few toponyms from the Spanish side of the border (Fig. 8.1).

A particular case is Sheet No. 1, corresponding to the international section of the River Minho, where the first preparatory work for the signing of the Treaty of Lisbon (Treaty on Boundaries, 1864) took place. An initial joint commission, composed of military officers and diplomats from both Iberian countries, conducted the first reconnaissance missions between 1855 and 1857, with the aim of preparing the demarcation of the border, resolving the main frontier disputes—even if only provisionally—and regulating the use of the river's water. However, as the Portuguese government did not accept all the conclusions, a second commission was convened between 1858 and 1862 to establish the points included in the treaty, particularly the division of the various islets in the river (García Álvarez and Puente Lozano 2017).

In any case, due to its historical and geographical particularities—being a natural boundary that served as a frontier even before Portugal's independence in the twelfth century, and marked by a distinctive flow and fluvial regime, along with the structures built over the centuries by riverside communities on both banks as expressions of their way of life (such as fishing weirs, jetties, docks and walls)—the border was not formally demarcated until 30 May 1897. It was recorded cartographically in a detailed atlas consisting of 59 sheets, at a scale of 1:2,500 (García Álvarez and Puente-Lozano 2015).

Sheet 1 of the *Carta chorographica de Portugal* depicts approximately eighty kilometres of the lower course of the River Minho (Baixo Miño from the Galician perspective, or Alto Minho from the Portuguese perspective), corresponding to the international section. Although fieldwork began as early as 1868, it continued until 1883, and the sheet was not published until 1887—ten years before the signing of the *Acta relativa à entrega do rio Minho* (the Proceeding Concerning the Handover of the River Minho Border). For this reason, the border line was not depicted.

In the representation of the Raia, a stark contrast is evident: while in the Minho valley a strip of Spanish territory approximately four kilometres wide is depicted, in the remaining border areas along the Trancoso and Castro Laboreiro rivers, the Spanish side appears as a blank space. It can be inferred that the rivers generally marked the border, but this was not the case for the River Minho, whose waters and river islets were the subject of ongoing disputes. As a result, particular care was given to the position and shape of the islets, with the main ones being named (García Álvarez and Puente-Lozano 2016).

In this mapped territory, the River Minho functions as an axis of symmetry along a cross-border strip approximately eight kilometres wide, with the same pattern of dispersed settlement, a road network that develops both parallel and perpendicular to the river, twin fortifications facing one another, and similar patterns of land use.

Other elements can also be identified that helped to unite the two riverbanks and reinforce the closeness between the *transminhota* (trans-Minho) communities, such as the newly built road-rail bridge linking Valença and Tui (inaugurated in 1886), the international railway itself—

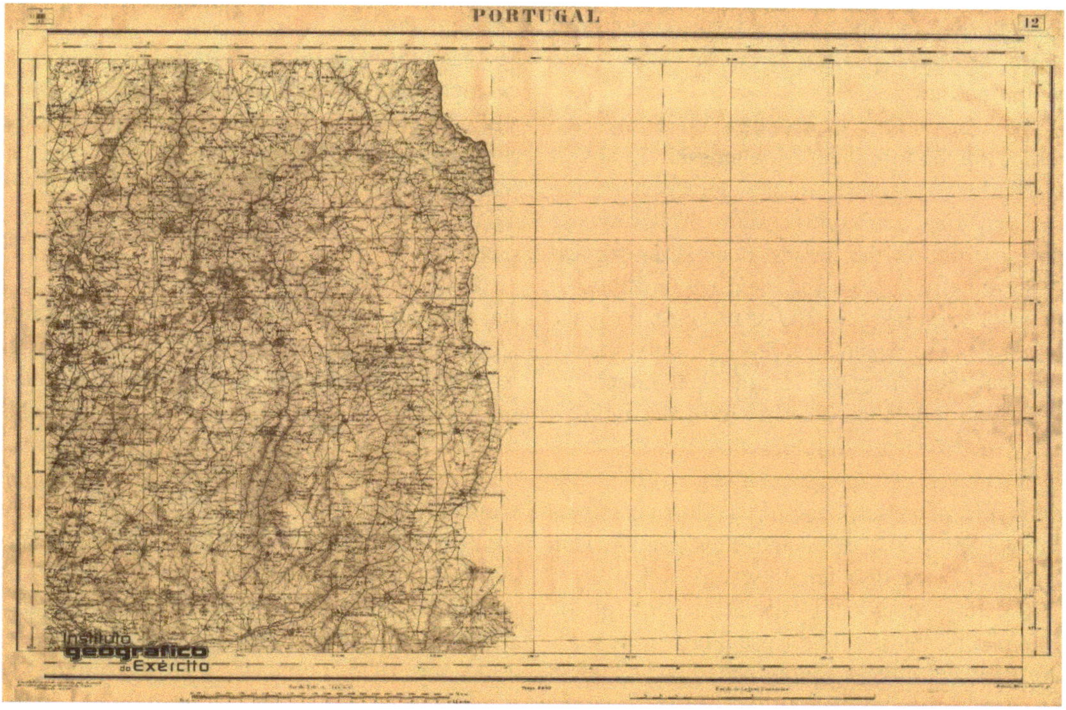

Fig. 8.1 [*Guarda, Carta corográfica de*] *Portugal*, Sheet 12, scale 1:100,000. Direcção-Geral dos Trabalhos Geodésicos do Reino (Directorate-General for Geodetic Works of the Kingdom), 1890. *Source* CIGeoE

which, even today, remains one of the few direct rail links between Portugal and Spain—as well as the various ferry crossings, which until then had been the main means of crossing the river and served as an indicator of the region's socio-economic vitality. Taken as a whole, the spatial coherence of the Minho River basin area accounts for the absence of a visible border line or any formal political division.

Another particular case is reflected in Sheets 3 and 6, published in 1896 and 1904 respectively. The area depicted corresponds to that described in the *Acta da entrega da linha de fronteira entre Portugal e Hespanha, desde o Penedo dos três reinos até ao rio Douro* (Proceedings of Handover of the Border Line between Portugal and Spain, from the Penedo dos Três Reinos to the River Douro), dated February 1901 (Martins 2002). The entire Raia Seca is marked by a dotted line which, in general, follows ridgelines or coincides with watercourses, seeking to align the political border with 'natural boundaries'—albeit with some exceptions. The values indicated along the border line correspond to elevation points (altitude in metres) rather than the numbers of the border markers (a total of 145 in this section), although in some stretches, certain symbols appear to represent border markers. As for the Raia Húmida, while the border symbol is drawn over the River Mente, this is not the case for the River Maçãs or the River Douro.

On the other side of the line, there is an almost total absence of information; only on Sheet 6, a few toponyms of borderland settlements on the Spanish side are recorded.

While the representation of the northern border did not pose any major difficulty—largely because the demarcation work was already well advanced when the chorographic survey of this part of the country was carried out—the opposite was true in the south, where some sheets were published before the joint commission had begun its fieldwork. This is the case with Sheet

26, published in 1866, precisely the same year in which the annexes to the Treaty of Lisbon were being finalised.

At first glance, it becomes clear that the border line on this sheet is either entirely absent or nearly so. Indeed, in the depiction of the southern area, the configuration of Portuguese territory follows the natural boundaries formed by the Ribeira do Caia and the River Guadiana—in stark contrast to the vast empty space shown on the Spanish side. The only exceptions are the locations of the settlements of 'Badajoz' and 'Villa Real', within the municipality of Olivença—although this disputed territory receives no reference–along with a few toponyms on the left bank of the Guadiana, indicating river ports, mills and elevation markers related to Portuguese territory.

The representation of the Raia north of the Ribeira do Caia is more diffuse. Portuguese territory does not always follow watercourses (as in the case of the River Xévora), nor the ridgeline, suggesting an area of uncertain or as yet undefined jurisdiction—a point underscored by the prominence of the toponym *Referta* (dispute in Portuguese, referring to its contested status).

However, it should be emphasised that this map series does not constitute a unified image of the Luso-Spanish border. Rather, it is a set of juxtaposed cartographic fragments depicting this space—stretching from the mouth of the River Minho to the mouth of the River Guadiana—as it was portrayed over the second half of the nineteenth century. It remained unchanged when reissued in 1944.

8.3.2 The *Carta Itinerária De Portugal,* 1:250,000

This analysis is based on a copy of the *Carta itinerária* published in 12 sheets, presumably issued sometime between 1903 and 1921, although the field surveys were carried out in the final decade of the nineteenth century. As the title suggests, this is an itinerary map intended to meet the needs of the army, enabling the planning or anticipation of troop movements and the development of a defence strategy in the event of a land invasion. In terms of the representation of routes, three types are indicated: *estrada a macadam* (macadamised road), shown as a solid red line; *estrada a macadam em construcção* (macadamised road under construction), shown as a dashed red line; and *caminhos* (paths), represented by a solid green line.

The entire so-called border line is drawn as a sequential line composed of small crosses. There are no longer any unclear territories or zones of uncertain sovereignty—not even in the case of Olivença, where the map aligns the political border with the River Guadiana, nor in the undefined areas of the *refertas* of Alegrete and Ouguela. The border south of the confluence of the Ribeira de Cuncos with the River Guadiana, continuing to its mouth, is also already shown as delimited, even though the reconnaissance and demarcation process only took place between 1904 and 1926, culminating in the signing of the *Convênio de Limites* (Convention of Limits) (Santos Sánchez 2023) (Fig. 8.2).

Generally, only Portuguese territory is represented, and almost all routes come to a halt at the border line—in stark contrast to the vast cartographic silence on the Spanish side, which is devoid of labels. However, a few signs of cross-border connections can be observed, but only on the sheets covering the northern frontier. For example, on Sheet 1, corresponding to the districts of Viana do Castelo and Braga, not only are the border crossing points indicated—such as those at Valença and Lindoso—but some nearby routes and settlements are also shown.

Sheet 2, covering the province of Trás-os-Montes (districts of Vila Real and Bragança), presents a distinctive depiction of the border: the representation of the territory does not stop at the demarcation line; instead, the roads and paths link Portuguese settlements to neighbouring Spanish localities and to one another, forming a local road network.

In turn, the detailed depiction of the Raia's hydrographic network appears to suggest that, between the valleys of the Tâmega and Douro rivers, political boundaries do not always correspond to natural features—particularly the

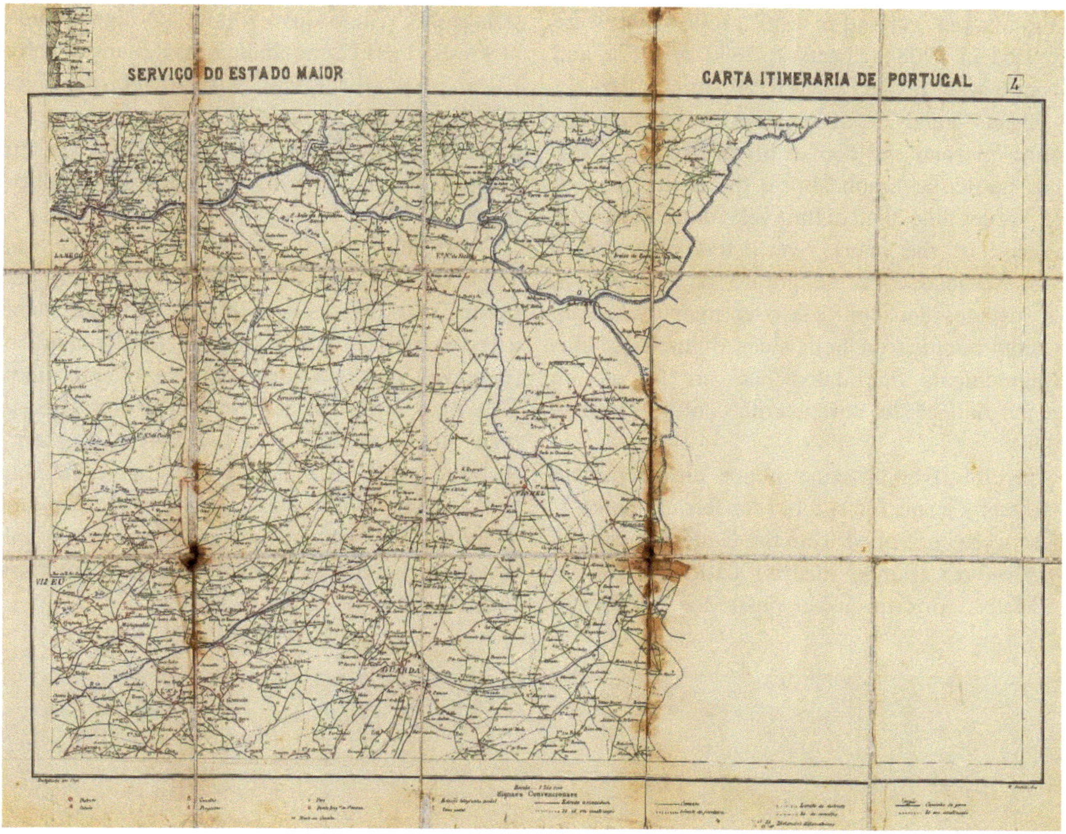

Fig. 8.2 [*Guarda*] *Carta itinerária de Portugal*, Sheet 4, scale 1:250,000. Serviço do Estado-Maior, 1905. *Source* CIGeoE, B1HC-G1-07

region's orography and topography—since the ridgeline rarely serves as the political border.

8.3.3 The *Carta Chorographica dos Limites de Fronteira*, 1:50,000

The particular features of this map—especially the fact that it was surveyed at a topographic scale and depicts a symmetrical strip of territory on both sides of the border—make it the most suitable for establishing a 'portrait' of the Raia. Above all, it avoids the imbalance in the representation of the two territories evident in the two previously analysed examples.

Given that this is a scientific document—geodetic and topographic in nature—with simultaneous political, diplomatic and military significance, surveyed (as previously mentioned) by the International Border Commission between Portugal and Spain and validated by the governments of both Iberian countries, one would expect a uniform and scientifically 'neutral' depiction of the border, enabling an unambiguous reading of the territory.

The process by which this map was produced is not fully known. In fact, none of the 22 published sheets identify the coordinating or publishing body, nor the place of publication. In 1904, a new, revised and updated, version of the map was prepared based on the first edition of 1893, which appears to have had limited circulation.

The production of this new edition may have continued over the following years, since Sheet

13 is marked 'revised in 1907', four others were 'revised in 1908' (Sheets 1, 2, 11 and 12), and Sheet 8 was 'revised in 1910' (Garcia 2000).

Sheets 1 and 2 of this printed series chart the entire international course of the River Minho, with particular emphasis on the delineation of the border line that definitively determined the division of the many river islets downstream from Monção. This sought to put an end to the disputes and conflicts over their ownership by communities on both sides of the border—disagreements that dated back as far as the second half of the seventeenth century (Herzog 2018).

Several river crossing points are identified, marked by a specific symbol for ferries. Three of these were controlled from the Galician bank, by *carabineiros* (border guards) stationed in Vilar de Matos, Amorim and opposite the Portuguese village of Cristelo Covo. On the Portuguese side, however, there is no indication of any Guarda Fiscal posts (the Fiscal Guards were the force responsible for monitoring border crossings)—despite the fact that they were already there at the time and present in significant numbers (Santos 1985).

When analysing the representation of the Raia Seca, the border line highlights the 'artificial border markers (stone pillars)', indicated with a red symbol, and the 'natural border markers (engraved cross)', shown with a blue symbol. The map legend states that 'the numbers in black ink correspond to the border markers'. A final note explains that 'the remaining conventions follow those of the *Carta chorographica de Portugal*', suggesting a direct link between these two contemporaneous projects (Fig. 8.3).

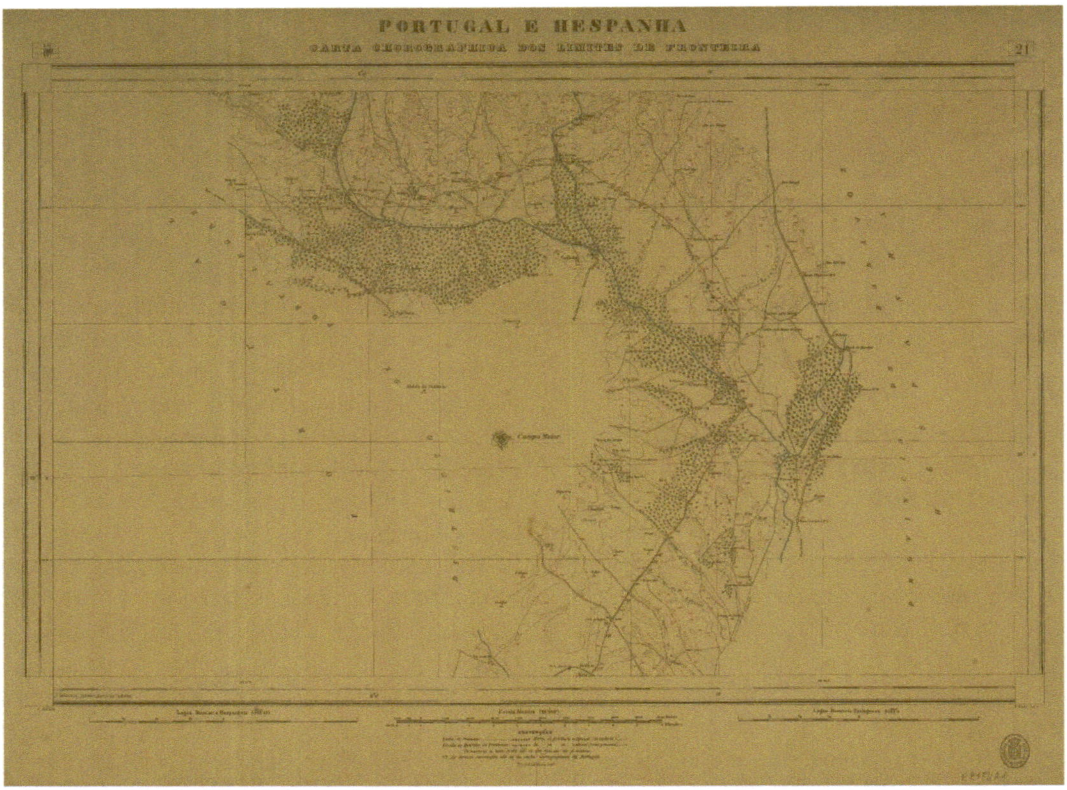

Fig. 8.3 *Portugal e Hespanha. Carta chorographica dos limites de fronteira* (Portugal and Spain. Chorographic Map of the Border Limits), Sheet 21, Scale 1:50,000. [Comissão Internacional de Limites], undated. *Source* CIGeoE, t24-C-R0150

An analysis of Sheet 6, relating to the land border between northern Portugal (Trás-os-Montes) and Galicia, shows the Raia between border marker no. 272, near São Vicente da Raia, and marker no. 378, at the 'Malhada do Salgueiral', near the Spanish village of Tejera/Teixeira. Almost all these markers are artificial, and several of them are identified—in addition to some natural ones.

The organisation of the borderland area further underscores the artificial nature of the political dividing line. Indeed, the distribution of settlement clusters and the dense network of paths connecting them reveal a shared territory, shaped by communities with a common agro-pastoral way of life and a subsistence economy based on agricultural production in the valley floors and the use of highland pastures.

The contrast between this section of the border and other parts of the Trás-os-Montes frontier could not be more striking: while the four-kilometre strip of the Portuguese Raia contains thirty-five settlements—including hamlets and villages, with a comparable number on the Spanish side—the northeastern sector, represented on Sheet 7, is far more sparsely populated, with only eight Portuguese and around a dozen Spanish villages. This disparity may be explained by geological composition, soil type and the relative abundance of water (Taborda 1987).

The dense network of narrow, difficult-to-navigate paths is only interrupted by the steep valley of the River Mente, which could be forded at certain marked crossing points—such as the Puerto de Espiño and the Porto de Lodeiros. Even so, it facilitated significant circulation between neighbouring settlements, a pattern that the demarcation line did not substantially disrupt, at least in the early years. One need only recall the events of the so-called 'monarchist incursion' of 1911, which took place in this very border region (Moreira 2019).

It is therefore unsurprising that the Treaty of Lisbon mentions the existence of various tracts of land disputed by the populations of these borderland villages, referring in Article XXVIII to the existence of 'common roads, with free passage of livestock for the peoples of both kingdoms' (Martins 2002). In terms of the presence of state authority, only one barracks belonging to the *carabineiros* is marked—in the Galician village of Berrande.

In turn, Sheet 21, depicting the border strip between the Alto Alentejo and the Spanish province of Badajoz, already records the resolution and division of the *refertas* of Arronches and Ouguela in the Alentejo, as agreed by both parties in article XXII of the Treaty of Lisbon. The border line was defined by a series of artificial markers, numbered 714–795. Once again, the political division does not align with either the area's orography or hydrography.

Here too, numerous roads and paths crossed the demarcation line, having long served as routes for local populations to access communal pastures.

Given the particular features of this territory, the presence of the state is more visible in the landscape, evident in the three Guarda Fiscal posts on the Portuguese side and two huts belonging to the *carabineros* on the Spanish side.

8.4 Conclusion

An analysis of the various cartographic examples shows that the representation of the Raia varies not only for technical reasons—such as map typology, survey methods, cartographic scales and the institutions responsible—but also due to political, diplomatic and military considerations.

In the case of the *Carta chorographica de Portugal*, at a scale of 1:100,000, the map was produced by military technicians under the authority of a civil institution—although, given its chorographic features, it also served military interests. The *Carta itinerária de Portugal*, by contrast, was a military project with numerous civil applications (notably as a road map); in both cases, the border line, even when not depicted, nonetheless aims to establish a clear division or separation between territories.

It should be emphasised that scientifically rigorous representation of the territory's main topographic and hydrographic features—including the entire border strip—was the principal aim of Portuguese cartography. Thus, the decision to leave the Spanish territory blank, without any form of identification or labelling, was clearly a political one. It served to reinforce the separatist nature of the border line, marking the distinction between the material, cartographic evidence for 'us' and the vast cartographic silence regarding the 'other'. In this regard, both maps show an effort to align the topographical and hydrographical depiction of the territory with political boundaries, as does the socio-economic representation, which breaks off at the 'line'.

There appears to be a clear geopolitical intention at play here, especially when we consider that the long process of establishing the border coincided with major political events that gave rise to two opposing currents of political, cultural and economic thought. On the one hand were the 'Iberianist' positions (championed by both monarchists and federalist republicans, particularly in Spain); on the other, their opposition, which found expression in nationalist claims on the Portuguese side—both by the state itself and by the republican party, who viewed Spain as a constant threat and 'danger', and as the primary reference point in shaping Portuguese national identity, constructed in opposition (Meireles 2017). In this sense, the attempt to erase the rest of the peninsula from view was meant to emphasise the national character of the territory.

It is worth noting that these two cartographic collections, in addition to being publicly accessible, were also part of the pedagogical materials supplied to the Liceus de Portugal—the Portuguese state secondary schools. In this way, the construction of national identity and the nation-state was shaped by the cartographic images that adorned the classroom walls: beyond the border line, there was only emptiness.

However, the silencing of Spanish territory also extended to Olivença. Although the Portuguese government continued to claim full sovereignty over that municipality, in practice the issue had disappeared from the political agenda by the second half of the nineteenth century, only regaining brief and limited momentum following the establishment of the republican regime in 1910. In effect, both maps presented the political border as coinciding with the River Guadiana—thus excluding Olivença from Portuguese territory.

The cartographic analysis of the Luso-Spanish border presented in the *Carta chorographica dos limites de fronteira* is more balanced and equitable, despite having been produced using the Portuguese geodetic network and the conventions of the *Carta chorographica de Portugal* at a scale of 1:100,000. In fact, there is a marked concern to ensure a consistent and balanced representation of the territory. Thus, in addition to the four-kilometre-wide strip on each side of the border line—showing topography, elevation, hydrography, settlement patterns and road networks—the remaining space is not an absolute void: some settlements are named, and the adjacent provinces (on the Spanish side) and districts (on the Portuguese side) are identified. Only on Sheet 21 (Campo Maior) are the names of the two Iberian countries explicitly indicated.

The map reveals a deeply interconnected borderland—one that appears far more transboundary than divided or isolated—over which a dividing line was imposed. However, this line, at least initially, does not seem to disrupt the existing spatial organisation, particularly given the still-limited presence of the state apparatus, especially on the Portuguese side.

The discourse of scientific objectivity underpinning the delineation of the border line—as fixed in the various maps analysed here and used as the basis for national cartographic representations—served to legitimise the state's official geopolitical decisions.

In other words, cartographic identity, as fixed through the maps, helped shape national identity around an anti-Iberianist and anti-Spanish stance (the silence on the maps) which the republican governments that came to power after 1910 particularly sought to assert, using the country's African colonial empire to validate its position on the international stage.

The very act of delimiting and demarcating the border, and of representing the territory cartographically, constituted a form of geopolitical discourse, which the ruling elites harnessed to shape the political consciousness of the nation (Sidaway 2000).

Archival Sources

CIGeoE. Centro de Informação Geoespacial do Exército (Army Geospatial Information Centre), Lisboa.

References

Acuerdos fronterizos con Portugal y Francia. Colección de recopilaciones (2005). Ministerio de Asuntos Exteriores y Cooperación, Madrid, vol 1

Alegria MF, Daveau S, Garcia J, Relaño F (2007) Portuguese cartography in the Renaissance. In: Woodward D (coord) The history of cartography. Vol. III—cartography in the renaissance, Part 1. University of Chicago Press, Chicago-London, pp 975–1068

Biggs M (1999) Putting the state on the map: cartography, territory, and European state formation. Comp Stud Soc Hist 41(2):374–411. https://doi.org/10.1017/S0010417599002080

Branch J (2010) "Colonial reflection" and territoriality: the peripheral origins of sovereign statehood. Eur J Int Rel 18(2):277–297

Branco R (2003) O mapa de Portugal: Estado, território e poder no Portugal de Oitocentos. Livros Horizonte, Lisboa

Brotton J (2019) História do mundo em 12 mapas. Edições 70, Lisboa

Cairo H, Godinho P, Pereiro X (coords) (2009) Portugal e Espanha: Entre discursos de centro e práticas de fronteira. Colibri, Lisboa

Catroga F (2013) A geografia dos afetos pátrios. Edições Almedina, Lisboa

Conrad B, Křížová M (2023) Borders in early modern history (ca. 1500–1800). In: Hansen J et al (eds) The European experience: a multi-perspective history of modern Europe, 1500–2000. Open Book Publishers, pp 37–44

Cunha L (2007) Linhas cartográficas no fio do tempo: Notas para uma navegação ibérica. In: Cunha M, Cunha L (orgs), Intersecções ibéricas. Margens, passagens e fronteiras (pp 15–33). 90º Graus, Lisboa

Dias MH (2003) As explorações geográficas dos finais de Setecentos e a grande aventura da Carta Geral do Reino de Portugal. Revista da Faculdade de Letras – Geografia I (XIX): 383–396

Dias MH (2009) Finis Portugalliae: Nos confins de Portugal. Instituto Geográfico do Exército, Lisboa

Dias MH (2018) A longa gestação da Carta Itinerária Militar de Portugal. Boletim Do Centro De Informação Geoespacial Do Exército 80:16–39

Dias MH, Ferreira Botelho H (1999) Quatro séculos de imagens da cartografia portuguesa (2ª ed.) Comissão Nacional de Geografia, Centro de Estudos Geográficos, Instituto Geográfico do Exército, Lisboa

Facal JL (2013) Breve historia cultural de los nacionalismos europeos. Catarata, Madrid

Ferreira M (2001) O Tratado de Madrid e o Brasil Meridional: Os trabalhos demarcadores das partidas do Sul e a sua produção cartográfica (1749–1761). Comissão Nacional para as Comemorações dos Descobrimentos Portugueses, Lisboa

Garcia J (1996) A configuração da fronteira luso-espanhola nos mapas dos séculos XV a XVIII. Treballs de la Societat Catalana de Geografia XI(41):293–321

Garcia J (2000) A fronteira impressa: Apontamentos sobre uma série cartográfica. In López-Trigal L, Guichard F (eds) La frontera hispano-portuguesa: Nuevo espacio de atracción y cooperación. Fundación Rei Afonso Henriques, Zamora, pp 57–72

García-Álvarez J (2013) Los mapas. In Moreno J, Núñez XM (eds), Ser españoles. Imaginarios nacionalistas en el siglo XX. RBA, Barcelona, pp 315–363

García-Álvarez J, Puente-Lozano P (2015) La delimitación de la frontera luso-española en el río Miño (1855–1906): Conflictos territoriales y representaciones cartográficas. Terra Brasilis (Nova Série) 6. http://terrabrasilis.revues.org/1708

García-Álvarez J, Puente-Lozano P (2016) La frontera hispano-portuguesa en el río Miño en el Tratado de Límites de 1864: Deslinde y ordenación de un território fluvial conflictivo. In: Vera F, Olcina J, Hernández M (eds) Paisaje, cultura territorial y vivencia de la geografía. Libro homenaje al profesor Alfredo Morales Gil. Universidad de Alicante, Alicante, pp 761–787. https://doi.org/10.14198/LibroHomenajeAlfredoMorales2016-34

García-Álvarez J, Puente-Lozano P (2017) A controversial natural border: the making of the Spanish-Portuguese boundary along the Minho River (1855–1866). Geoforum 79:5–16

García-Álvarez J, Santos-Sánchez C (2023) La frontera hispano-portuguesa: Proceso de formación y conflictos territoriales. Salamanca. Revista De Estudios 67:15–35

Goettlich K (2019) The rise of linear borders in world politics. Eur J Int Rel 25(1):203–228

Hagen J, Diener A (2012) Borders: a very short introduction. Oxford University Press, Oxford

Halmos K, Marin I, Masař T (2023) Borders in modern history (1800–1900). In: Hansen J et al (eds) The European experience: a multi-perspective history of modern Europe, 1500–2000. Open Book Publishers, pp 45–52

Harley JB (1988a) Maps, knowledge and power. In: Cosgrove D, Daniels S (eds) The iconography of landscape: essays on the symbolic representation, design and use of past environments. Cambridge University Press, Cambridge, pp 277–312

Harley JB (1988b) Silences and secrecy: the hidden agenda of cartography in early modern Europe. Imago Mundi 40(1):57–76. https://doi.org/10.1080/03085698808592639

Herzog T (2018) Fronteiras da posse: Portugal e Espanha na Europa e na América. Imprensa de Ciências Sociais, Lisboa

Hobsbawm E, Ranger T (1988) The invention of tradition. Cambridge University Press, Cambridge

Konvitz J (1987) Cartography in France (1660–1848): science, engineering, and statecraft. University of Chicago Press, Chicago

Magalhães JR, Garcia J, Flores J (coords) (1997) Cartografia e diplomacia no Brasil do século XVIII. Comissão Nacional para as Comemorações dos Descobrimentos Portugueses, Lisboa

Martins M (2002) Delimitação da linha de fronteira entre Trás-os-Montes e Castela-Leão, em 1901. População e Sociedade 8:321–365

Meireles M (2017) Do iberismo ao ibero-americanismo, da federação latina à confederação luso-brasileira – percursos e reflexões de alguns intelectuais portugueses entre os séculos XIX e XX. Intellèctus 16(2):23–44. https://doi.org/10.12957/intellectus.2017.31650

Moreira LM (2019) A República e a Monarquia em confronto: a guerra civil portuguesa na raia galega (1911–1912). Revista de Historiografía 16(30):97–118. https://doi.org/10.20318/revhisto.2019.4745

Nadal F, Urteaga, L (1990) Cartography and state: national topographic maps and territorial statistics in the nineteenth century, Geo-Critica. English Parallel Series, no. 2, pp 9–67

Palmeirim AX (1845) Memoria sobre topographia portuguêza. Revista Universal Lisbonense, V 1:55

Pelletier M (2019) Carte de France. In: Edney MH, Pedley MS (eds), The history of cartography. Vol. IV—Cartography in the European Enlightenment. University of Chicago Press, Chicago-London, pp 39–244

Pinto S, Comissão de Negócios Estrangeiros e Comunidades Portuguesas (2022) Olivença na história. Assembleia da República, Lisboa

Pro Ruiz J (2007) A concepção política do território e a construção do Estado espanhol: Cartografia, cadastro e administração. In Almeida O, Branco R (coords), Burocracia, Estado e território. Portugal e Espanha (séculos XIX-XX). Livros Horizonte, Lisboa, pp 183–202

Santos P (1985) Génese e estrutura da Guarda Fiscal (Ensaio histórico). Imprensa Nacional-Casa da Moeda, Lisboa

Sidaway JD (2000) Iberian geopolitics. In: Dodds K, Atkinson D (eds) Geopolitical traditions: a century of geopolitical thought. Routledge, London, pp 118–149

Silva AC (1998) O modelo espacial do Estado moderno: Reorganização territorial em Portugal nos finais do Antigo Regime. Editorial Estampa, Lisboa

Santos-Sánchez C (2022) Border disputes in Southern Portugal and Spain in the nineteenth and twentieth centuries. Finisterra 57(120):23–44

Santos-Sánchez C (2023) La construcción de la frontera moderna: El proceso de elaboración del Convenio de Límites de 1926 entre España y Portugal (PhD thesis), Universidad Carlos III de Madrid, Madrid

Stoian S (2014) The establishment and demarcation of borders in Europe in the early modern age. Research and Science Today, Supplement 2, 6–15. https://www.proquest.com/scholarly-journals/establishment-demarcation-borders-europe-early/docview/1614660902/se-2

Taborda V (1987) Alto Trás-os-Montes: Estudo geográfico. Livros Horizonte, Lisboa

Vandenbroucke M, D'heygere BW (2021) Het Verdrag van Kortrijk 1820: Geschiedenis van de Belgish-Franse grens van De Panne tot Luxemburg. Mens En Cultuur Uitgevers, Ghent

Withers C (2001) Geography, science and national identity: Scotland since 1520. Cambridge University Press, Cambridge

Wintle M (2016) Emergent nationalism in European maps of the eighteenth century. In Jensen L (ed) The roots of nationalism: National identity formation in early modern Europe, 1600–1815. Amsterdam University Press, Amsterdam, pp 271–288. https://www.jstor.org/stable/j.ctv8pzcpr.17

Kolibri 360 GmbH
Ferdinand-Jühlke-Straße 7
99095 Erfurt, Deutschland
produktsicherheit@kolibri360.de